Patrick Moore's

PASSION FOR

ASTRONOMY

Patrick Moore's

PASSION FOR

ASTRONOMY

David & Charles

Photographs by the author unless credited otherwise

Frontispiece: Saturn, 1990 – from the Hubble Space Telescope. This was the first good planetary picure received from the space telescope

A note on metrication

To turn inches into millimetres, multiply by 25.4
To turn inches into centimetres, multiply by 2.54
To turn feet into metres, multiply by 0.305
To turn miles into kilometres, multiply by 1.609
To turn millimetres into inches, multiply by 0.039
To turn centimetres into inches, multiply by 0.394
To turn metres into feet, multiply by 3.281

Temperatures.
If F= degrees Fahrenheit and C= degrees Celsius:
F= (C x 1.8) + 32
C= (F -32) ÷ 1.8

A catalogue record for this book is available from the British Library.

ISBN 0 7153 0182 9

First published in hardback 1991
Published in paperback 1993

Printed in Portugal by Resopal
for David & Charles plc
Brunel House Newton Abbot Devon

Contents

Under the Stars 6
Starting Out 18
'The Most Interesting Place on Earth' 27
Moonshine 44
Neighbours 62
Voyaging 76
Tales of the Unexpected 95
Astronomers in Strange Haunts 110
The Lives of the Stars 122
Star Legends and Sky Gods 134
The Sky from Pole to Pole 146
Who Can Hear Us? 160
Island Universes 172
The End of Time 185
Glossary 195
Useful Addresses 201
Acknowledgements 202
Index 203

Under the Stars

Why is a starlit sky so fascinating? It is only too easy to reply 'Because it is there', but the reason goes much deeper than that. The stars make us realise that we on Earth are only a very small part of a universe which is so vast that we cannot begin to understand it. At first glance there seem to be millions of stars on view, and on a dark, clear night the sky looks so crowded that finding one's way around appears to be a daunting task. In fact it is quite easy; nobody can ever see more than about two and a half thousand stars at any one

The Earth from space. Despite the clouds, the outline of Africa is very obvious

The Ptolemaic system, with the Earth lying at rest in the centre of the universe

time without using optical aid, and making yourself familiar with the main patterns or constellations takes only a few evenings. It is a worthwhile exercise, as I found out at the early age of six. The stars become so much more interesting when you learn how to tell which is which.

Of course, we have the advantage of many centuries of scientific progress, and our remote ancestors had ideas very different from ours — which was natural enough. To them the Earth was flat, and lay at rest in the exact centre of the universe, with everything revolving round it once in twenty-four hours. The stars were tiny lamps, fastened on to an invisible crystal sphere. With hindsight it is easy to laugh at these ideas, but originally there was nothing absurd about them; after all, the Earth does look flat apart from local hills and dips, and the sky does indeed appear to turn round, taking the Sun, Moon and stars with it. It was also quite reasonable to believe that the celestial orbs had been installed specially for our benefit, either by a collection of deities of various types and dispositions or else (in later religions) by a single all-powerful God. Now and again, signs of

A small Sun?

Heraclitus of Ephesus, who was born about 544BC, believed the diameter of the Sun to be 1ft.

divine displeasure were seen; eclipses, auroræ, showers of meteors, and — above all — comets, which were taken to be omens of evil. Remember the lines in Shakespeare's *Julius Cæsar:* 'When beggars die, there are no comets seen; the heavens themselves blaze forth the death of princes.'

Modern astronomers are quick to stress how insignificant we are. The Earth is a small planet, moving round the Sun, which is a very ordinary dwarf star, and is only one of an assembly of at least a hundred thousand million stars making up what we call the Galaxy. The Galaxy itself is part of a whole universe of similar systems, some of which are a great deal larger and more populous than ours. Moreover, most astronomers believe that life is widespread in the universe, and that our particular form of it is likely to be very lowly. But the star-gazers of thousands of years ago would not have agreed. Their teaching was not to degrade our importance, but to magnify it, and to question this fundamental principle was regarded as heresy.

Heretics emerged now and then, and were vocal; one of the earliest, the Greek philosopher Anaxagoras, was exiled from his home in Athens for daring to teach that the Sun is a red-hot stone larger than the Peloponnesus. In enlightened Christian times the fate of heretics was much worse. In 1600, Giordano Bruno was burned at the stake in Rome, one of his 'crimes' being his belief that the Earth moves round the Sun instead of vice versa. It was only in the nineteenth century that Roman Catholics were officially allowed to read the great book by Galileo in which the Earth was reduced to the status of a normal planet.

Three major steps had to be taken before there could be any real understanding of the nature of the universe. First, the idea of a flat Earth had to be given up, and this was achieved in Greek times by

The shape of the Earth

Over the years there have been many different ideas about the shape of the world. The ancient Egyptians believed the Earth to be in the form of a rectangular box, with the sky being formed by a goddess with the rather appropriate name of Nut. According to some Indian tribes, on the other hand, the Earth is carried on the body of an elephant, which is in its turn resting on the shell of a turtle floating in a limitless sea; one is bound to think unkindly of turtle soup. Both St Augustine and (later) Martin Luther insisted that the Earth had to be flat, as otherwise people living on the underside would be unable to witness Christ's descent on Judgement Day.

The Full Moon. There are no shadows, and even large craters are difficult to identify

Eclipse of the Moon, 13 March 1979 (Commander H.R. Hatfield)

sheer observation. Aristotle, who lived around 350BC, pointed out that the view of the sky changes according to one's position on the surface of the Earth, so that, for instance, the brilliant star Canopus can be seen from Alexandria but not from Athens. And when the Earth's shadow falls upon the Moon, during a lunar eclipse, the shadow is curved, showing that the Earth's surface must also be curved. Aristotle gave other proofs as well, but these two were enough.

Step No 2, to dethrone the Earth from its proud central position, was much more difficult. A few of the Greeks made the attempt, but unfortunately they could give no proofs, and they found very few followers. It was not until less than five hundred years ago that astronomers began to change their minds. To give the whole story would be out of place here; suffice to say that the revolution in outlook was more or less complete by the time that Isaac Newton published his great book about gravitation in 1687.

Step No 3 involved appreciating the nature of the stars themselves, and to realise that their distances are inconceivably vast. Finding the distance of a star is not easy, to put it mildly, and it defeated all the efforts of astronomers up to the year 1838, when it was solved by Friedrich Wilhelm Bessel by means of a method known as parallax — basically the same as that used by a surveyor who wants to measure the distance of some inaccessible object such as a mountain-top. Bessel found that a dim star, known only by its catalogue number

Moon in eclipse

In the first edition of his famous novel *King Solomon's Mines,* H. Rider Haggard described a full moon, an eclipse of the Sun, and another full moon in quick succession. When somconc pointed out that there had been a slight mistake he prudently replaced his solar eclipse with an eclipse of the Moon.

PHASES OF THE MOON

The Moon has a diameter of 2,158 miles, and is therefore much smaller than the Earth. Its orbital period is 27.3217 days.

Because it has no light of its own, it shows regular phases, or apparent changes of shape, from new to full. When it is almost between the Earth and the Sun, its dark side is turned toward us, and we cannot see it at all; this is the true 'new moon', though the term is often applied to the narrow crescent seen soon afterwards in the evening sky just after sunset. The crescent thickens, until the Moon has become half; this is known (rather confusingly) as First Quarter, because the Moon has then completed one-quarter of its orbit. The phase then becomes gibbous (between half and full), and then full, after which the phases repeat themselves in the reverse order: gibbous, half (Last Quarter) and back to new.

Though the Moon's orbital period is 27.3 days, both it and the Earth are moving round the Sun, and this means that the interval between one new moon and the next (or one full moon and the next) is 29 days 12hr 44mins — officially termed the synodic period. It is therefore possible for February to lack a new moon or a full moon.

The hollow Earth

Captain John Cleves Symmes, of the United States Army, was convinced that the Earth was hollow, and that inside it were five concentric spheres, each of which was inhabited. In 1823 he tried to persuade Congress to send an expedition to the North Pole, where he believed that there was an entrance to the inner regions. Congress refused, but it is worth noting that no less than twenty-five members voted in favour of Symmes' resolution.

(Many years later, William Gladstone, Prime Minister of Britain, wanted the Treasury to finance an expedition to the Atlantic in search of the sunken continent of Atlantis. He, too, met with a polite but firm rebuff.)

The Crescent Moon, photographed with a 12½in reflector by the late Henry Brinton

The Half Moon.
The 'seas' are clear, 15in reflector

of 61 Cygni, lies at a distance of around 60 million million miles. Light, travelling at a rate of 186,000 miles per second, takes eleven years to cover this distance, so we say that 61 Cygni is eleven 'light-years' away.

Most of the other stars are much more remote than this; 61 Cygni ranks as a near neighbour. It follows that when we look at them, we are seeing them not as they are now, but as they used to be in the past. Light takes 1¼ seconds to reach us from the Moon, which is much the closest natural object in the sky; 8.6 minutes from the Sun; just over four years from the nearest bright star beyond the Sun, Alpha Centauri; 680 years from the Pole Star, 1,800 years from Deneb in the constellation of the Swan, 22,500 years from the cluster of stars in Hercules, and so on. Thus in 1990 we see the Moon and Sun in their present-day form, but the light now reaching us from the Pole Star began its journey toward us at the time of the Crusades, while we see Deneb as it used to be when the Romans occupied Britain. I will have more to say about this later, but it is always worth bearing in mind that once we look beyond the Solar System, our immediate part of the universe, our information is always bound to be very out of date. In saying this, of course, we are referring to the sort of knowledge drawn from physical experience. Whether 'thought' travels instantaneously, for example, is quite another matter, and this is something which scientists cannot tell us — at least, not yet.

If the stars are so far away, then they must also be very powerful.

Strange experiment

One of the most remarkable scientific experiments of modern times was carried out at Magdeburg, in Germany, in 1933. Rocket enthusiasts, including Wernher von Braun (later to design the V2 weapons, and also to master-mind the first American artificial satellite) were making their early tests, but were handicapped by lack of funds. Fortunately for them, a member of the Magdeburg City Council had become convinced that the Earth's surface was the inside of a hollow sphere, so that a rocket launched upward from Europe would land somewhere in the Antipodes. Accordingly, the City Council provided money for von Braun's team to make a practical test. In fact two rockets were launched, but the first one attained a maximum altitude of 6ft and the second took off horizontally instead of vertically, so that the results were inconclusive.

Deneb is the equal of at least 70,000 Suns put together, and even Deneb is by no means exceptional, though certainly it ranks as a 'supergiant'. On the other hand, we must not be too humble. The Sun may be classed as a dwarf, but there are many stars which are much less luminous. If we represent the Sun by a pocket torch, other stars will range from glow-worms up to brilliant searchlights.

The stars differ not only in power and distance, but also in colour. I remember that this was one of the first points I noticed when I started to find my way around. In a way I was lucky, because by using my outline star-map, bought with my pocket money at the high cost of sixpence, I managed to pick out Orion, which is one of the most striking of all the constellations and which dominates the evening sky all through winter and early spring. My map told me that Orion contained two exceptionally brilliant stars, one in the upper left corner of the main pattern and the other in the lower right; their names, I found, were Betelgeux and Rigel. To my surprise I saw that

Opposite: Scorpius. The Scorpion, a brilliant constellation; only part of it is visible from Britain. The leading star is the red Antares. This photograph was taken from South Africa, where the entire constellation is visible

THE EARTH IN THE UNIVERSE

It is a long time now since mankind believed the Earth to be the most important body in the universe. In fact it is a very insignificant member of the Sun's family, and seems important to us only because it is our home in space.

The Sun is a star, and a very ordinary star at that; astronomers even relegate it to the status of a stellar dwarf. It has a diameter of around 865,000 miles, and lies at an average distance of 93,000,000 miles from us. In everyday life this may seem a long way, but it is not far to an astronomer, who has to become used to vast distances and immense spans of time.

Round the Sun move the nine known *planets*. Mercury and Venus are closer to the Sun than we are; then comes the Earth, and then Mars. Beyond Mars there is a wide gap, in which move thousands of dwarf worlds known as minor planets or *asteroids*. Next come the four giant planets: Jupiter, Saturn, Uranus and Neptune, plus another small world, Pluto, which does not seem to fit into the general scheme, and may not be worthy of true planetary status.

Most of the planets have secondary bodies or *satellites* moving round them. The Earth has one satellite: the Moon. Saturn has 18 known satellites, Jupiter 16, Uranus 15, Neptune 8, Mars 2 and Pluto 1, so that only Mercury and Venus are moonless.

The *Solar System* is completed by various bodies of lesser importance: *comets*, which are made up of small icy nuclei together with gas and dust, and smaller particles known collectively as *meteoroids*, some of which may dash into the Earth's upper air and show up as what we usually call shooting-stars. The average shooting-star meteor is smaller than a pin's head.

The members of the Sun's family have no light of their own, and shine only by reflecting the rays of the Sun. This does not apply to the *stars*, of course, because the stars are suns in their own right — some of them a great deal larger, hotter and more powerful than ours. They appear so much smaller and fainter only because they are so much further away. Even the nearest star beyond the Sun lies at a distance of about 25 million million miles.

In our particular star-system or *Galaxy* there are approximately 100,000 million stars. The system is a flattened one, with the Sun lying well away from the centre but almost in the main plane. This means that when we look along the 'thickness' of the system we see many stars in almost the same direction, and it is this which produces the appearance of the Milky Way in the night sky.

Beyond our Galaxy there are other *galaxies*, so far away that few of them can be seen without optical aid. From Britain, the most conspicuous of these outer systems is the Andromeda Galaxy, whose distance from us is of the order of 12 million million million miles. Though it is considerably larger than our own Galaxy, it can only just be seen with the naked eye as a dim, misty patch.

Our telescopes can show us thousands of millions of galaxies, so that the universe is a very large place indeed. No doubt many of the stars in our Galaxy and others are attended by inhabited planets, but as yet we have no proof.

Opposite:
Figure of Orion. The Celestial Hunter; from an old star map

while Rigel is pure white, Betelgeux is orange-red. This was my first introduction to the study of stellar evolution.

The colour of a star depends upon its temperature. Rigel is white, the Sun is yellow and Betelgeux is orange-red; therefore Rigel must be hotter than the Sun, while Betelgeux is cooler. To compensate for this low temperature, Betelgeux is extremely large. Its diameter is of the order of 250,000,000 miles, so that it could quite easily swallow up the entire orbit of the Earth round the Sun.

To the newcomer, whether young or old, revelations of this kind come as quite a shock, and once the initial awe has worn off it is time to look at the starlit sky with something more than casual interest. My own method of identifying the constellations was to pick out a few which could hardly be missed, and then use them as 'pointers' to the less obvious groups. Things are made much easier by the fact that, to all intents and purposes, the stars do not seem to move relative to each other; if we could board a time machine and go back to the time of King Canute, or even Homer, we would see the same constellations as those of today. The stars are not really fixed in space, and are moving about in all sorts of directions at all sorts of speeds, but they are so remote that their individual or 'proper' motions are too small to be noticed by the naked-eye observer even over periods of many lifetimes. Not so with our neighbours, the members of the Sun's family, which shift quite markedly even over short periods. In fact,

Ursa Major. The Great Bear, with its seven leading stars forming the pattern of the Plough or Dipper

ORION
Pl.29.
GEMINI
TAURUS
MONOCEROS
LEPUS
ERIDANUS
E
W
S
Meissa
Betelgeux
Bellatrix
Mintaka
Anilam
Alnitak
Thabit
Rigel
Saiph
Sidy. Hall, sculp.

ASTROLOGY

It is strange to find that even today there are still many followers of astrology, which may be called the superstition of the stars. Until a few centuries ago astrology was regarded as a true science, but there is no excuse for it today.

The cult is based entirely upon the apparent positions of the Sun, Moon and planets against the background of stars. The stars are so remote that to all intents and purposes the star-patterns or constellations remain unchanged, but the members of the Solar System wander slowly around from one constellation to another, keeping to a well-defined band in the sky known as the Zodiac. Astrologers claim that a person's character and destiny is influenced, or even controlled, by the positions of the planets at the time of his (or her) birth.

In fact, constellation patterns have no real significance, because the stars in any particular constellation are at very different distances from us, and again we are dealing with nothing more significant than line of sight effects. Moreover, the names of the constellations are purely arbitrary, and mean nothing at all. To give just one example: the stars making up what we call Capricornus, the Sea-goat, are spread out in a more or less formless way, and the outline bears not the slightest resemblance to a goat, marine or otherwise.

If you hold up your hand against a background of clouds, it is absurd to claim that your hand is 'in' the clouds. It is therefore equally absurd to claim that a planet seen against the background of stars in a constellation is 'in' that constellation.

Astrologers in general do not try to explain 'why' astrology works; they merely say that it 'does'. Actually, what they are doing is coincidence-hunting, and if you make enough predictions some of them will turn out to be right; to quote Judge Stephen, 'it is impossible always to be wrong' (though some modern politicians do their best). The best that can be said about astrology is that it is fairly harmless so long as it is confined to circus tents, seaside pier-heads, and the columns of the less serious Sunday papers.

STELLAR PARALLAX

The first man to publish his successful measurement of the distance of a star was Friedrich Wilhelm Bessel, Director of the Königsberg Observatory, in 1838. His method was that of trigonometrical parallax.

A good demonstration to explain the principle is to shut one eye, hold out a finger at arm's length, and align the finger with a relatively distant object such as a clock on the wall. Now, without moving your face or your finger, use the other eye. Your finger will no longer be aligned with the clock, because you are observing from a slightly different direction; your two eyes are not in the same place. If you know the distance between you eyes, and also the amount of angular shift of your finger against its background, you can 'solve the triangle' and work out the distance between your finger and your face.

This is the method used by a surveyor to measure the distance of an inaccessible object such as a mountain-top. Of course, a very long baseline is needed to measure the angular shift or *parallax* of any star. Bessel therefore selected a star (61 Cygni) which he correctly believed to be one of our nearer neighbours, and observed it at an interval of six months, during which time the Earth had moved from one side of its orbit to the other. Bessel knew that the Earth's distance from the Sun was 93,000,000 miles; therefore the diameter of his baseline was twice this, or 186,000,000 miles, and the parallax of 61 Cygni against the background of more remote stars was detectable, even though it amounted to less than one-third of a second of arc. At about the same time similar results were obtained elsewhere; Thomas Henderson measured the distance of the bright southern star Alpha Centauri, while F. Struve, in Russia, obtained a rather less accurate result for Vega in the constellation of Lyra.

The parallax method can be used for close stars, but beyond a distance of a few hundreds of light-years the shifts become so small that they are swamped by unavoidable errors of observation, and less direct methods have to be used, most of which involve finding the star's real luminosity and then checking it against its apparent brightness. However, it was parallax which gave the first real indication of the true scale of the universe.

Messier 31, the Andromeda Galaxy (200in photograph). The nearest of the really large outer galaxies and, at a distance of 2,200,000 light years, the most remote object clearly visible with the naked eye (Palomar Observatory)

this was how the planets were first identified; the name means 'wanderer'.

Once you have learned the main groups, you will find that there is plenty of interest, even if you have no optical aid. Orange and red stars can be seen here and there; Betelgeux is one of the brightest examples, but there are others (Antares in the Scorpion has been said, rather luridly, to look like a glowing coal). Some stars are double, while others are grouped in clusters, and there are occasional misty patches which are called nebulæ, from the Latin word for 'clouds'. If you know just where to look, you can even see a remote star-system, the Andromeda Galaxy, which is so far away that its light takes over two million years to reach us.

Most people remember the nursery rhyme which begins 'Twinkle twinkle, little star . . . ' It is true that stars do seem to twinkle, but this has nothing directly to do with the stars themselves, and is due solely to the Earth's dirty, unsteady atmosphere through which the starlight has to pass before reaching us at ground level. The starlight is, so to speak, 'shaken about', and twinkling is the result. The effect is most pronounced with a star which is low over the horizon, because its light is coming to us through a much thicker layer of air, as you can see for yourself if you look first at a low-down star and then at a star of about the same brightness which is higher up. Planets do not twinkle to the same extent, because a planet appears as a small disk, whereas a star is to all intents and purposes a point source of light.

So much for our introduction. What next?

Parallax

If a star showed a parallax of one second of arc, it would lie at a distance of 3.26 light-years. This distance is therefore known as the parsec, and is generally used by professional astronomers in preference to the light-year.

Starting Out

> **An unexpected mirror**
>
> Telescope optics can sometimes be found in unexpected places. One of America's leading optical experts, A.G. Ingalls, was once staying in a small hotel in the heart of the country when he realised that there was something unusual about the shaving mirror in his bedroom. Since he always carried optical equipment around with him, he made a quick test, and found that he was confronted with a first-class 12in mirror for an astronomical reflector. Apparently he put in a prompt bid for it, though not before he had seen the hotel proprietor and explained his reasons.
>
> Since then I have always looked closely at hotel shaving mirrors, but up to now I have had no luck.

Anyone who has taken matters this far will be well on the way toward developing a 'passion for astronomy', but one or two notorious pitfalls lie close ahead, and this is a good time to say something about them, because they are of tremendous importance. Ignore them, and you may be in danger of losing your interest altogether — which, believe me, will be your loss.

It is often believed that to become an astronomer, even an amateur one, you must have a telescope, and telescopes which are of any real use are expensive items. It is not worth attempting to give prices here, because they change rapidly according to the current rate of inflation. When I was a boy I bought a splendid telescope for £7.10s; I still have it, and still use it. This, however, was in 1929. Today, the same telescope would cost something of the order of £400, which seems exorbitant until you take the trouble to compare it with a couple of British Rail tickets between, say, London and Glasgow.

Basically, the thing *not* to do is to rush out to the nearest camera shop and pay several tens of pounds for a small, nice-looking telescope, usually Japanese. It will not serve you well, mainly because it is not large enough to collect a sufficient amount of light from objects in the sky.

Telescopes are of two types: refractors and reflectors. A refractor collects its light by means of a special lens known as an object-glass, while a reflector uses a curved mirror. The greater the size of the object-glass (for a refractor) or the main mirror (for a reflector), the greater the light-grasp. Astronomically, the minimum useful size is probably 3in for a refractor and 6in for an ordinary reflector. Smaller telescopes are often unsatisfactory, though I would be the first to concede that they are better than nothing at all.

Bear in mind that though the initial cost of a useful telescope seems high, it will provide you with something which will last you for a lifetime provided that you take reasonable care of it. Extra equipment can be added at leisure; it is the basic instrument which is all-important.

Unfortunately, great care is needed when buying a telescope, even a new one. Everything depends upon the quality of the optics. If the

This rotating observatory, made and used at Selsey by the late Reg Spry, looks remarkably like an ordinary shed. The entire structure revolves, rather than just the upper part

Odds and ends

Tubes for reflectors can be made in all sorts of ways. Selsey amateur, the late Reg Spry, who wrote a splendid little book about astronomical ways of using 'odds and ends', used a very convenient 6in reflector whose 'tube' was made out of two dismantled larder shelves.

object-glass or mirror is poor, then the performance also will be poor, and faulty optics do not always betray themselves at first glance. If you can enlist the aid of a friendly expert, then so much the better; your local astronomical society may well be able to help.

In any case, it is wrong to say that every astronomer must have a telescope. There is plenty to be seen with the naked eye. Moreover, there is an excellent solution available to the would-be observer who cannot afford a good telescope, or has no site from which it could be used. Buy binoculars!

A pair of binoculars is made up of two small refracting telescopes linked together. They are classified according to the magnification obtained, and the diameter of each object-glass expressed in millimetres. Thus a pair of 7x50 binoculars means that the magnification is 7, and that each object-glass is 50mm across. Good binoculars have most of the advantages of very small telescopes, and

Night light

Amateur astronomers can often be ingenious. One friend of mine had trouble with the local Council, who insisted upon putting a street lamp in a position from which its rays would shine straight into his observatory. Having considered and rejected the idea of buying an air-gun, he went out in the dead of night, climbed the lamp post and painted the glass black. This was achieved some years ago, and to the best of my knowledge the Council has yet to realise that this particular lamp gives out no more radiance than a glow worm.

THE ASTRONOMICAL REFRACTOR

There is no basic difference between an 'everyday' telescope, used for bird-watching and similar hobbies, and an astronomical refractor. Both collect their light by means of a glass lens or object-glass, so that the light-rays are brought to focus and the image is enlarged by a second lens, known as an eyepiece. However, there are important differences in detail. It is true that a bird-watcher's telescope can give nice low-power views of the Moon and star-fields, but the more dedicated amateur astronomer will need something rather more tailored to his own interests.

First, an 'everyday' telescope will have a fixed magnification, usually low, and a wide field. Secondly, it is bound to be light enough to be hand-held, so that there is no chance of providing it with a more powerful eyepiece to give enlarged views of objects in the sky. With an astronomical telescope, on the other hand, there will be several eyepieces which can be interchanged. Note, in particular, that all the actual magnification is carried out by the eyepiece. It is the function of the object-glass to collect the light in the first place — and, of course, the larger the object-glass, the more light it can provide.

A favourite mistake by beginners is to attempt to use too high a magnification. It is a general rule that the maximum really useful power is a magnification of 50 (x50) per inch of aperture. Thus with a 3in refractor — that is to say, a telescope with an object-glass 3in across — there is no point in trying to use a magnification higher than 3 x 50, or 150. In theory you can use any power you like, but with every increase the resulting image becomes fainter. Use a power of, say, x500 on a 3in refractor, and the image will be so dim that you will be lucky to see it at all. Certainly there will be no chance of seeing detail on the Moon or anything else. If you want to use an eyepiece giving x500, you must obtain a larger telescope.

Refractors give delightfully sharp images; they are easy to use, and they need no maintenance apart from reasonable care. A 3in is probably the ideal instrument for the beginner. The main drawback is expense, and any refractor more than 4in in aperture is a very costly item indeed. It is also true that a 4in aperture is probably the limit for portability. If you aspire to anything larger, you will have to start thinking in terms of a permanent observatory.

Remember, too, that an astronomical telescope will give an inverted image. Everything will appear upside-down. In fact, any telescope will do this, but in an 'everyday' instrument an extra lens-system is incorporated to make the image erect. However, every time a beam of light is passed through a lens it is slightly weakened, and though this does not in the least matter when looking at a soaring bird, or observing a ship out at sea, it matters very much to the astronomer, who is anxious to collect all the light he can. Therefore, the extra lens system is left out, which is why most astronomical drawings and photographs of the Moon and planets are oriented with south at the top. It is quite possible to buy an extra erecting lens and use it on an astronomical telescope, but there is really very little point in doing so.

REFLECTING TELESCOPES

A reflecting telescope has no object-glass; the light is collected by a specially curved mirror, usually made of glass and coated with a thin layer of aluminium to make it reflect as much light as possible.

With the Newtonian reflector, the light comes down an open tube, strikes the main mirror or speculum, and is reflected back up the tube on to a smaller, flat mirror placed at an angle of 45°. This secondary mirror directs the light-rays into the side of the tube, where an image is formed and is magnified by an eyepiece in the usual way. With a Newtonian, then, the observer looks 'into' the tube instead of up it. The tube need not be solid; a skeleton framework will do quite well, because the only requirement is to hold the optics in the correct positions. As with a refractor, the maximum really useful magnification is x50 per inch of aperture.

A Newtonian reflector is cheaper than a refractor of equivalent size, but it is more temperamental, and is always liable to go out of adjustment; it is also worth bearing in mind that the mirrors need periodical re-coating. On the credit side, any really skilled amateur can make a good mirror without having to buy prohibitively expensive equipment, whereas lens-making is a task for the real expert.

With the Cassegrain reflector, the secondary mirror is convex rather than flat, and the light is directed back down the tube to the eyepiece via a hole in the main speculum. There are also various other patterns, some of which involve both mirrors and lenses, but on the whole it seems that the simple, straight-forward Newtonian is the best 'starter' for anyone who prefers a reflector to a refractor.

3in refractor on pillar and claw mounting. This is the same telescope as on page 25 but on its original mounting – which is not to be recommended!

few of the drawbacks, though it is true that the magnification is low and cannot be altered.

Binoculars are relatively cheap even if bought new, and there is a better chance of obtaining a second-hand bargain, because you can actually test the binoculars before buying them. If they do not focus properly, or if there is any false colour round a bright image, then look elsewhere. With a magnification greater than about x10, the binoculars become rather too heavy to be hand-held with comfort, and some sort of tripod or a neck mounting is needed, so that on the whole I would opt for something in the 7x50 to 8x50 range — though opinions differ, and everyone is bound to have his own ideas.

Now let us go back under our starlit sky. Your binoculars will bring out the different star colours well, and stars which seem white with the naked eye may turn out to be orange or red. Find some double stars, such as Mizar in the famous constellation of the Great Bear, which never sets over Britain and is therefore always available whenever the sky is clear and dark. Look for any star-clusters which are above the horizon, and search around for nebulæ, which we know to be stellar nurseries — places where fresh stars are being formed out of the gas and dust spread thinly through space. You may even pick up some galaxies, millions of light-years away. They will not look spectacular, but at least you can satisfy yourself that they really exist.

What of the members of the Sun's family or Solar System, made up of the Moon, planets and the rest? The Moon is a joy. I admit to being prejudiced, because at an early stage, as soon as I had learned the fundamentals and decided to make astronomy my main hobby, I had my first view of the lunar mountains and craters through a

Spode's Law

According to what astronomers call Spode's Law ('if things *can* be awkward, they *are*), a tall tree always lies in the most inconvenient part of the landscape. If the offender is on one's own property, of course, something can be done. For example, adjoining my main observatory was a tall, stately pear tree which never gave any pears, and became an unmitigated nuisance. One night it obscured my view of Saturn. Next day it turned into a small, stumpy pear tree, and since then it has provided plenty of pears each year, so that obviously the message was received and understood.

If the tree is not one's own, things are more difficult. If the owner is not sympathetic, the only solution is to make the telescope portable. One frustrated amateur has mounted his 8in telescope upon a wheelbarrow; it looks peculiar, but it does solve the problem.

Opposite:
The author's main telescope; a 15in reflector with an octagonal wooden tube and a massive fork mounting, set up at Selsey

friend's telescope, and from that moment onward the Moon became my favourite object; it still is. Even with the naked eye you can see the dark patches which are mis-called 'seas', and binoculars bring out so much detail that you may wonder how anyone could map it. Also, the view of the Moon's surface changes from night to night, almost from hour to hour, as the Sun rises or sets over the lunar surface. When the Moon appears as a crescent, or a half, look for the shadow-filled craters; near full, the scene is dominated by the bright streaks or rays which spread out across the surface for hundreds of miles. Armed with an outline map, you can soon learn your way around.

The planets are less suited to the binocular-owner, because to see them properly you do need more magnification than normal binoculars can give; but you should be able to make out the changing appearance of Venus, and perhaps the four large moons of Jupiter. For Saturn's rings, and the snowy caps of Mars, you will have to invest in a telescope.

Night view of the author's second dome, containing an 8½in reflector at Selsey

FULLERSCOPES

TELESCOPE MOUNTINGS

Mountings are all-important; trying to use a good telescope on a shaky stand is no better than trying to use a good record-player with a bad needle. Everything has to be rock-steady, and the telescope must be capable of being moved smoothly and easily.

There are two types of mountings. With the 'altazimuth', the telescope can swing freely either in altitude or in an east-or-west direction *(azimuth);* for a very small telescope, a simple tripod will be quite acceptable. If you want to use higher magnifications, an equatorial mounting is highly desirable; here, the telescope is set up on an axis which is parallel to the Earth's axis of rotation, so that when the instrument is moved in azimuth the up-or-down motion will look after itself. Adding a mechanical drive means that the target object can be kept firmly in the field of view without the need for continual adjustment.

There are many designs for equatorial mountings, among them the 'German', where the telescope is balanced by a counterweight on the opposite side of the polar axis, and the 'fork', where the telescope is mounted between two prongs. All these can be made by anyone who is reasonably good with his hands, but there is one piece of advice which is well worth passing on. Work out the maximum weight you need to make the mounting really steady. When you have done so, multiply by three!

Beard on fire!

If you are projecting the Sun, make sure that your finder telescope is capped. An old friend of mine, a well-known solar observer, once forgot this — and found to his consternation that his beard was on fire!

Projecting the Sun. The image is sent on to a screen. The telescope is the author's small 4in Maksutov, a type of telescope using both lenses and mirrors

There is only one danger in astronomy (apart from the possibility of falling off an observatory roof, as has been known to happen!). This involves the Sun. Dwarf star though it may be, the Sun is very hot by everyday standards, and even its surface is at a temperature not far short of 6,000°C, while near its core, where its energy is being produced, the temperature rises to a staggering 14,000,000°C and perhaps rather more. In a way, the Sun may be likened to a controlled H-bomb, and even at its distance of 93,000,000 miles it must always be treated with respect.

Consider what will happen if you turn a telescope sunward, and then look through the eyepiece. You will focus the light, and the heat, on to your eye in the same way as a Boy Scout can start a camp fire by using a magnifying-glass on dry material such as straw. Look at the Sun in this way, and you will burn your eye so badly that permanent blindness will result. This is not mere alarmism; it has happened, and must be borne in mind. Do not look at the Sun even when a dark filter has been fitted over the end of the telescope. No filter can give full protection, quite apart from the fact that it is always liable to splinter so quickly that the luckless observer will have no time to remove his eye from the danger-zone.

Of course, there are ways of observing the Sun, and to see the changes in the darker patches or sunspots often visible on the disk, use your telescope as a projector by all means. But I can only repeat what I have said before literally thousands of times: there is only one rule for looking straight at the Sun — *don't*.

One factor today, which did not apply to the same extent before the

Pillar and claw

A well-known abomination is the pillar and claw stand, often found with small refractors. It looks nice, and is cheap, but is about as steady as the average blanc-mange. However, it is ideal for a small telescope which is to be stood in the corner of the library to look decorative, but is never actually turned toward the sky.

CHOOSING A TELESCOPE

Go to any camera shop, and you may well see small, nice-looking telescopes at what appear to be reasonable prices. A 2in refractor will run to less than £100 if mounted on a simple altazimuth tripod, no more than £200 if equatorial; a 4in Newtonian will be in the same range. They seem tempting, but my advice is, quite definitely, to avoid them. They will almost certainly be unsteady; they will have small fields of view; the reflectors at least will be difficult to set up and adjust, and they will have poor light-grasp. Admittedly, they are better than nothing at all — but given a choice between a very small telescope and a pair of binoculars, I would unhesitatingly opt for the binoculars, and save up for an adequate telescope — that is to say, a refractor with an aperture of 3in, or a reflector of at least 6in. (Inch for inch, a refractor is more effective than a Newtonian reflector.)

Second-hand telescopes can be obtained, and now and then one finds a real bargain, but there are many traps for the unwary. In particular, beware of misleading advertisements. One often finds telescopes advertised as 'up to 300 times magnification', or '10,000 area magnification', or something of the sort, without anything being said about the aperture. For example, I recently saw an advertisement for a telescope which, it was claimed, could give 'magnification times 400'. I made some inquiries, and found that the instrument was in fact a 2½in refractor. Using the basic rule, we find that the maximum useful power will be 2½ x 50 =125. Another favourite trick for disposing of a small refractor is to put a 'stop' inside the tube, below the object-glass, cutting out the defective part of the lens and reducing the aperture still further.

It is not really difficult to identify a poor refractor, but a reflector is more of a problem, because a bad mirror does not betray itself at first sight. The only real safeguard is to test it before handing over any money. Sadly, this does not apply only to second-hand telescopes, and it is sensible to take the greatest care; after all, you are buying something which will, hopefully, last you for many years.

war, is that there are many astronomical societies, both national and local. The advantages of joining one of these are enormous. You will be able to exchange ideas with other people of like interest; you will make many new friends, and you will no doubt find someone with a telescope. Later, if you like, you can even carry out some really useful scientific research.

There are many textbooks on astronomy. This is not one of them. All I want to do in the following chapters is to introduce you to some of the wonders which have enthralled me — and I hope that you, too, will enjoy yourselves among the stars.

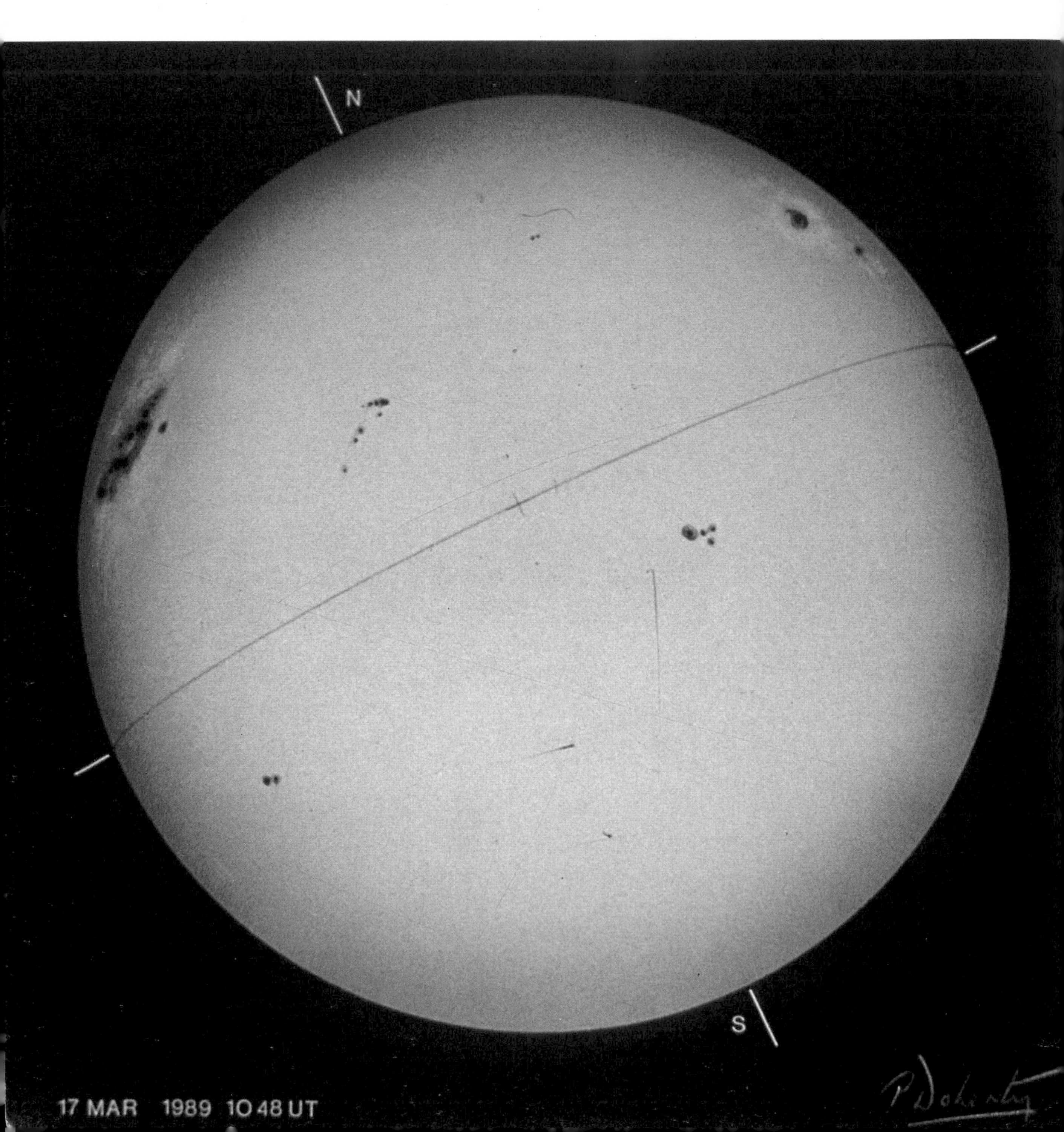

Great Sunspot, March 1989. This was the spot associated with the great aurora of 13 March. Drawing by Paul Doherty

'The Most Interesting Place on Earth'

Some years ago I found myself standing alone in a deserted place. There was no breath of wind; the Sun had gone down, and the first of the stars was starting to show against the dark blue of the sky. To all sides nothing could be seen except the slopes of a curved wall. I was standing inside Meteor Crater in Arizona, once described by Svante Arrhenius, the Swedish scientist whose work was good enough to win him a Nobel Prize, as 'the most interesting place on earth'.

On this occasion I was entirely on my own, because we were in the process of making a television programme about the crater, and all our equipment had been lifted down by helicopter; the idea was for the 'chopper' to fly over and take a picture of me walking across the crater floor. It worked very well, but there was a period of around an hour when I felt cut off from mankind. Nothing much grows at the bottom of the crater. There is a certain amount of scrubby vegetation, but that is all. The only signs of human habitation are represented by the remains of mineworks, left by earlier investigators who hoped to find a mass of valuable metal buried beneath the crater floor.

You can reach Meteor Crater easily enough. It lies not far from the town of Winslow, and it is an easy drive from there. Come along the celebrated Highway 99 until you reach the turning; simply drive on, and after a reasonable period you will come to the crater. Not that you will recognise it until you are almost on top of it, because although it is nearly 600ft deep there is not much in the way of a wall above the outer landscape. Like most impact craters, it is a sunken formation.

The name is wrong. It really should be Meteor*ite* Crater, because it was produced not by a particle of the shooting-star variety, but by a large mass of iron-rich material which swung around the Solar System for thousands of millions of years before it happened to collide with Planet Earth. This happened a long time ago — at least 22,000 years, and, according to recent estimates, considerably more. At any rate, there was nobody in the area to watch it fall. We do not know when *Homo sapiens* first came to Arizona, but it was not until Meteor Crater was already old. This is why the various legends about it must be ruled out of court.

Meteorite swords

Meteorites have been known for centuries, though their nature was not appreciated. The Sacred Stone at Mecca is definitely a meteorite; in India, the Emperor Jahangir ordered two sword-blades, a dagger and a knife to be made from the meteorite which fell at Jalandhar on 10 April 1621, and in the nineteenth century part of a South African meteorite was used to make a sword for the Emperor Alexander of Russia.

SHOOTING STARS

Most people have seen meteors, or shooting-stars, at one time or another — but not everybody knows just what they are! Certainly there is no connection between them and the real stars, and neither is there any close association between meteors, which are cometary débris, and crater-forming meteorites, which are much larger bodies and which probably come from the asteroid belt.

A meteor is a tiny particle moving round the Sun in the same manner as a planet. Since most meteors are no larger than sand-grains, they cannot be seen until they dash into the Earth's upper air, when they become heated by friction against the air-particles, and burn away in the familiar luminous streaks. What we actually see is not the meteor itself, but the effects which it produces as it plunges downward. It is moving at a very great speed, and may make its entry at anything up to 45 miles per second. It does not survive the complete drop to the ground, but 'burns out' at a height of 40 miles or so, and finishes its journey in the form of very fine 'dust'.

They are of two types: sporadic, and shower. Sporadic meteors may come from any part of the sky at any moment, and there is no way of predicting them. Shower meteors, on the other hand, are associated with 'parent' comets, so that even though there is no way of predicting individual meteors we do at least know when definite showers are due. Each time the Earth ploughs through the 'dusty trail' left by the comet there is a display of shooting-stars, and this happens many time in every year.

Each shower seems to radiate from one special point in the sky, which is appropriately termed the 'radiant' and is named after the constellation in which it lies; thus the August meteors, which are particularly spectacular, issue from Perseus and are therefore called the Perseids. The reason for this is that the meteors of the shower are moving through space in parallel paths, and what we are seeing is purely a perspective effect, just as anyone standing on a bridge overlooking a regular motorway will see that the lanes of the motorway seem to radiate from a point near the horizon (assuming, of course, that the motorway is straight).

Generally speaking a shower will last for a period of several nights, or even in some cases for several weeks, with a fairly well-defined maximum. The richness is measured by what is termed the ZHR, or Zenithal Hourly Rate, defined as the number of naked-eye meteors which would be expected to be seen by an observer under ideal conditions, with the radiant at the zenith or overhead point. In practice, of course, these conditions are never met, so that the actual observed rate will always be less than the theoretical ZHR.

Of the parent comets (see table opposite), all have known periods (only 3.3 years in the case of Encke) though that of Thatcher's Comet is very long; it was seen for the only time in 1861, and may be expected once more in or near the year 2276. (To prevent any possible misunderstanding, its discoverer — the American astronomer A.E. Thatcher — has no connection with any present-day politician!) The shower of early January, which has a short, sharp maximum, radiates from a position in the constellation of Boötes, and derives its name from a constellation called Quadrans (the Quadrant) which has lost its identity on modern maps.

Some of these showers are fairly consistent, because they are comparatively old, and the débris has had enough time to spread all round the orbit of the parent comet. This is the case with the Perseids, whose parent comet, discovered in 1862 independently by Lewis Swift and Horace Tuttle, has a period of 130 years, and last returned in 1992. Other showers are much more erratic, notably the Leonids of November. The parent comet, Tempel-Tuttle, has a period of approximately 33 years, and was last seen in 1965; its meteors are 'bunched up', and we pass through the main swarm only occasionally. Thus there were brilliant displays in 1799, 1833, 1866, and in 1966, when for a while the hourly rate exceeded 60,000; but in most years the ZHR is so low that the shower is barely detectable. We may hope for another 'firework show' from the Leonids in 1999.

Meteors are much more plentiful than might be thought, and it has been estimated that 75,000,000 of them above magnitude 5 enter the atmosphere daily, but of course their combined mass is very small. Amateurs can carry out valuable work in observing them, though it must be admitted that meteor patrols, either visual or photographic, can be extremely taxing — particularly on a cold winter's night.

There are also vast numbers of much smaller particles whose mass is too low for them to produce luminous effects when they dash into the upper air. They are known, rather misleadingly, as micrometeorites.

The Leonids, taken at the height of the display from Arizona in 1966 (H. Gee)

It has been claimed that the local Indians have a story that long ago there was 'a ball of fire from the sky' which scorched the whole landscape, and was centred upon the present site of Meteor Crater. It is certainly true that some of the Indians remain in awe of the place, and are reluctant to go there, particularly at night, when I agree that it can look decidedly eerie. But the whole time-scale is wrong. Though there is bound to be some uncertainty about the Crater's age, it cannot be younger than 20,000 years at least, and we can be sure that there were no Indians in Arizona as long ago as that.

In a way, it is lucky that the meteorite came down where it did. There are other impact craters on Earth (and plenty on the Moon, if you follow conventional theory), but none is as well preserved as that

The main annual showers are as follows:

Name	Begins	Max	Ends	Max ZHR	Parent comet
Quadrantids	1 Jan	3 Jan	6 Jan	110	
Lyrids	19 Apr	22 Apr	24 Apr	12	Thatcher
Eta Aquarids	2 May	4 May	7 May	20	Halley
Delta Aquarids	15 Jly	28 Jly	15 Aug	35	
Perseids	25 Jly	12 Aug	18 Aug	68	Swift-Tuttle
Draconids	10 Oct	10 Oct	10 Oct	var	Giacobini-Zinner
Orionids	16 Oct	21 Oct	26 Oct	30	Halley
Taurids	20 Oct	4 Nov	25 Nov	12	Encke
Leonids	15 Nov	17 Nov	19 Nov	var	Tempel-Tuttle
Geminids	7 Dec	14 Dec	15 Dec		
Ursids	17 Dec	22 Dec	24 Dec	12	Tuttle

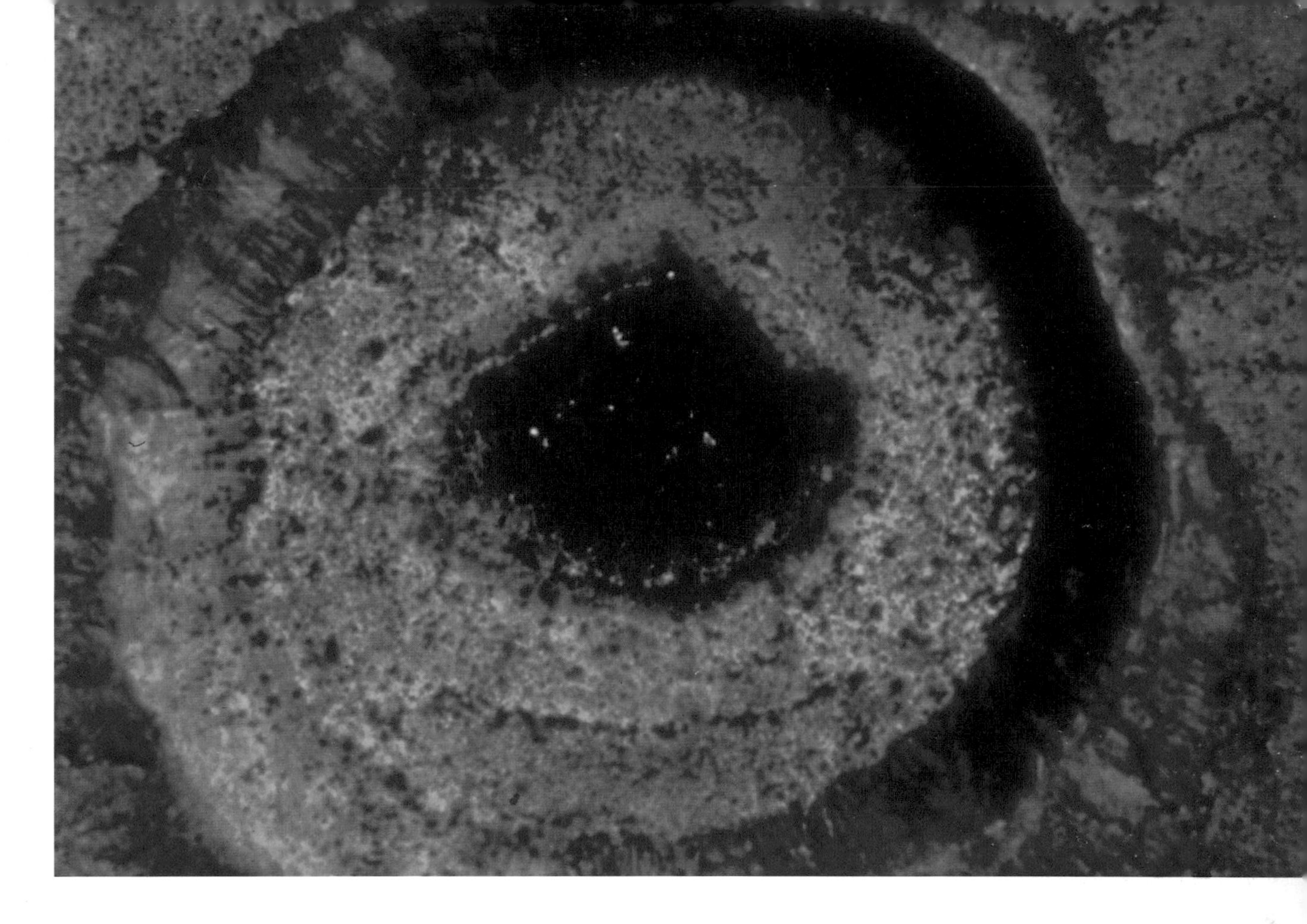

off Highway 99. There are two main reasons for this. First, the area is 'hard', and exceptionally well suited to preserving a huge scar; there is not much in the way of erosion, though no doubt the Crater has been filled up to some extent by windblown material and is less deep now than it used to be when first formed. Secondly, human activity has been at a low level. There has been only one attempt at serious interference with the Crater, and this piece of scientific vandalism was halted before it could do lasting damage.

Because the Crater is so isolated, it was not known to white men until little more than a century ago. Even when it was first reported, in 1871, nobody took much notice of it until the arrival of an American geologist, G. K. Gilbert, who was interested in it, surveyed it, wrote a report about it — and came to a completely wrong conclusion. He believed that the Crater was 'a steam explosion of volcanic origin'. (This is all the more significant because it was Gilbert who championed the theory that the craters of the Moon are of impact origin rather than being volcanic. Personally I believe that Gilbert was wrong there too, though I admit that I am in the minority.)

Next in the story of the Crater came Daniel Moreau Barringer, who plays the dual rôle of hero and villain. He looked at the Crater, and as a mining engineer he too was interested, though his reasons were very different from Gilbert's. Barringer seems to have realised quite quickly that the Crater was not volcanic at all, but had been gouged out by something which came from Outer Space. This 'something'

Above:
Wolf Creek Crater, Australia. This is much smaller than the Arizona Crater, but is definitely an impact structure

Above left:
Old ironworks in Meteor Crater (1980). The author is standing by the remains of the mining equipment abandoned long ago

Far left:
Aerial view of Meteor Crater, Arizona

could only be a meteorite. If extra proof were needed, there were meteorites lying about in the area, and it was easy enough to collect them. This is no longer the case, because all the available specimens have been gathered up, some to be put on display in the museum now established on the Crater wall; but Barringer's doubts, if he had ever had any, were set at rest. It is strange that Gilbert, with his geologist's eye, should have missed something which ought to have been so obvious.

Barringer also knew that meteorites are valuable. Some are stones, but others are made up of excellent-quality iron, and there are also traces of exotic materials such as iridium. If the meteorite were buried beneath the floor of the Crater, it would be a profitable business to mine it and sell it. So in his hero's capacity, Barringer bought the site and preserved it for posterity. In his villain's rôle, he began to work out how to delve down and fish the meteorite up. Clearly it would be too large to be brought to the surface intact, but if it could be opened up, so to speak, it could be chipped away.

Because the Crater is circular, it seemed axiomatic that the meteorite would be found directly underneath the centre of the Crater floor. In fact, this is not so. When an impacting object strikes, even at an angle, it will always produce a circular crater. When it hits, its kinetic energy is converted into heat, and it acts as a very powerful explosive. As soon as the meteorite has buried itself, the explosion blasts out a circular crater, and this is what happened in Arizona; we now think that what is left of the meteorite is not under the centre of the floor at all, but is somewhere below the Crater's south wall.

Barringer brought down mining gear, and started to look. He had no luck. Quite apart from the fact that he was searching in the wrong place, the material of the floor was unexpectedly hard; there is sandstone down to an appreciable depth, and then a layer of what is called kaibab dolomite, which is exceedingly tough. Then he switched his attention to other areas, with no better result. Finally, at a depth of less than 2,000ft, the drill jammed. Nothing more could be done, and Barringer gave up. A few later sporadic efforts were just as fruitless, and all that is left today is a remnant of the old mining gear, which has rusted and is very much of an eyesore.

Later still, it was found that the silica round the Crater rim is of very high quality, and again the world of Big Business showed interest. There was a considerable amount of activity, and the silica miners became an unmitigated nuisance, but this time officialdom stepped in, and for once on the right side. In 1967 the Crater was declared a National Natural Landmark. All work on excavation or collecting was stopped permanently, and there were no more suggestions about trying to find what lay under the Crater's south wall or anywhere else. A Board of Administration was set up, and a museum established on the Crater wall, just below the rim. It contains specimens of meteorites found in the area, as well as detailed explanations of how the Crater was formed.

We are used to regarding Government officials as scientific vandals, but in this case they acted very properly. Few people can

Meteor myths

Meteors can be quite startling; I have seen several which have outshone the Moon for a few seconds, and it is not surprising that there are many old myths about them. To the Hindus, they were pieces of the severed body of the god Rahu; in Islam, they were missiles thrown by angels at the evil spirits trying to eavesdrop at the gates of Heaven, while on the other hand the Russians took them to be demons which were in the process of being forcibly ejected from the Pearly Gates. The great philosopher Pliny the Elder, who lived around AD75, believed that every person had his own special star; when that star fell out of the sky, the person died.

The Romans were convinced that a meteor shower was a sign of approaching bad weather, but this is quite untrue; meteors and meteorites are totally unconnected with the climate, and it is equally wrong to suppose that meteorites are likely to fall during thunderstorms.

The last English meteorites

On Christmas Eve, 1965, a meteorite shot across the country and broke up; pieces of it were scattered all round the Leicestershire village of Barwell. One fragment went through the open window of a house, and was later found nesting coyly in a vase of artificial flowers. The last English meteorite – so far – fell at Glatton in Cambridgeshire on 5 May 1991, some 70ft away from a retired civil servant, a Mr Pettifer, who had been doing some casual gardening. It weighed 767g, and made a shallow depression about three-quarters of an inch deep.

The Henbury Craters in Australia (Northern Territory) are ancient impact structures. Photograph taken in 1986

mind paying for the privilege of going there; it is an experience not to be missed.

On my first visit, which must have been around 1955, one could walk down the 'Trail' in the wall and reach the floor. Today the Trail has been closed, because it is unquestionably dangerous, and apparently there was a bad accident some years ago. Not that it matters to the causal visitor; the best vantage point is from the crest of the wall, and it does not take too long to walk all the way round, because the whole Crater is less than a mile in diameter. But I do agree that to be on the floor of the Crater alone, under a darkening sky, is impressive by any standards.

What of other impact craters? There are quite a number here and there. One is at Wolf Creek, in Western Australia, which is rather like a smaller version of the Arizona Crater. At Henbury, in the Northern Territory, you will find a whole group of craters, and there is another not far away at Boxhole, largely covered with trees.

Also in the Northern Territory there is Gosse's Bluff, which is very ancient indeed — at least 50,000 years old, and probably more. Unlike the Arizona Crater, it has been very much eroded. Traces of it are unmistakable, and it is a singularly beautiful area, but to get there means a long drive across very rough country with no proper roads; a four-wheeled drive vehicle is strongly recommended. There is the remnant of a central structure, and indications of the old walls, but not much else which can be immediately put down to a cosmic

Fireballs

On the evening of 9 February 1913, people living in parts of the United States and Canada saw a unique event. From north-west to south-east a procession of majestically slow fireballs passed over South Ontario, burning and crumbling as they went; some were extinguished, but others went on until they were lost to view. Nothing like it has ever been seen before or since. The objects have become known as the Cyrillids.

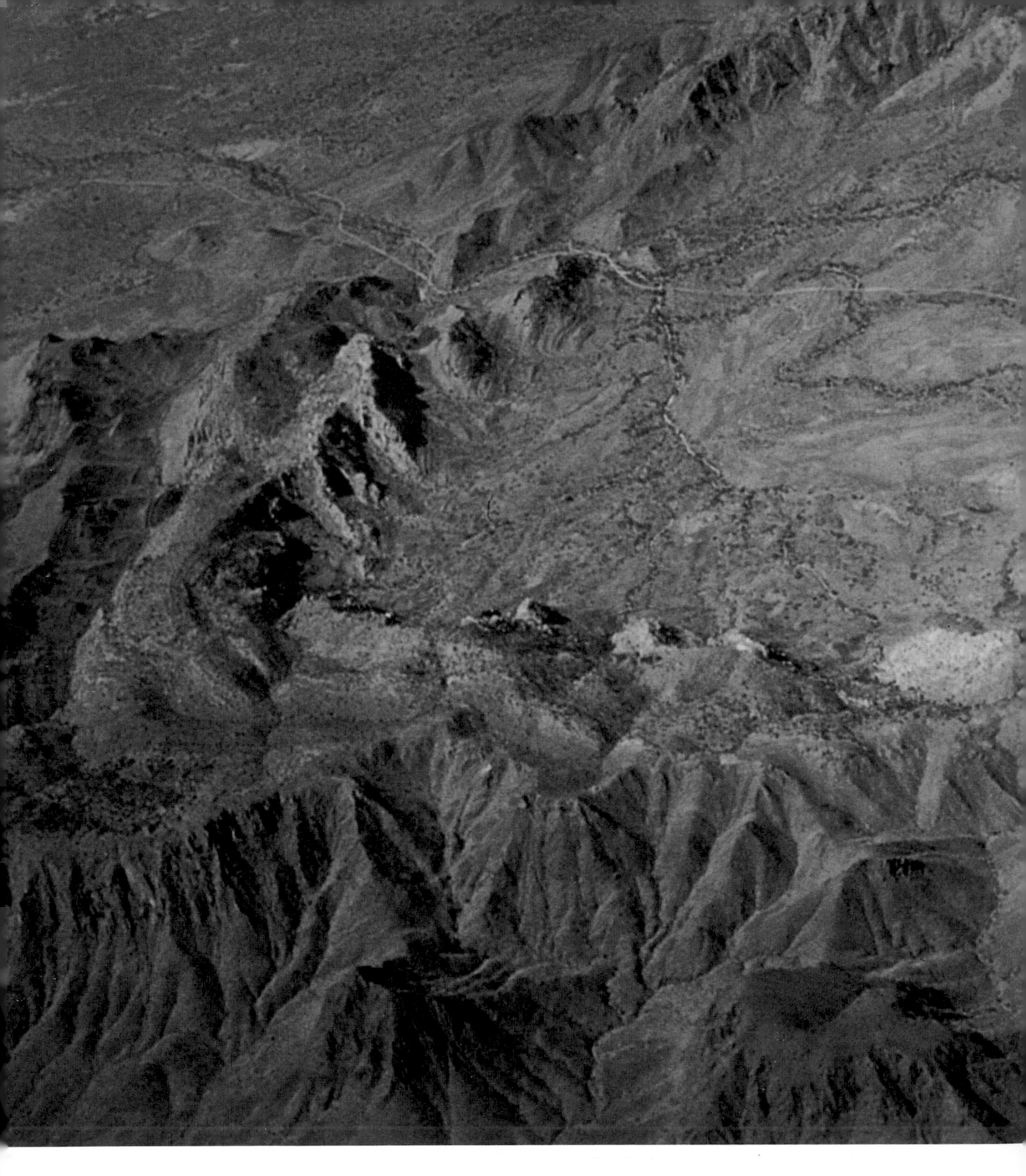

collision. Part of the region is a 'no go' area, because it is sacred to the Aborigines. It is unlikely to the highest degree that there were Aborigines in Northern Territory, or anywhere else, when Gosse's Bluff was formed; all the same, it is interesting, and I would not rule out any connection as firmly as I do of a link between the Arizona Crater and the legends of the American Indians.

Beware, however, of being too hasty about identifying impact

An aerial view of Gosse's Bluff in the Northern Territory of Australia. This is a very ancient impact structure (G. Gerrard)

Meteoroid visitor

On 19 August 1972 a meteoroid entered the Earth's atmosphere; it was first seen above Utah, when its height was later worked out to be 47 miles, and it reached its closest point to the ground above Montana, at 36 miles. It then began to move outward, and was last seen above Alberta at a height of 60 miles; it had been in view for a total of 1min 41sec, and at its best it was much brighter than the full moon. Its estimated diameter was 250ft, so that if it had landed in a populated area it would have caused a great deal of damage. This is the only known case of a meteoroid which has been seen to enter the atmosphere and then leave it again. Presumably the object is still in orbit round the Sun, and if we could now detect it we would certainly classify it as a small asteroid.

craters. Go to Pretoria in South Africa, drive north, and you will come to the Vredefort Ring, which is a hill-bounded depression containing two villages: Parys, and Vredefort itself. Unquestionably it is a crater, and many lists attribute it to impact. One 'confirmation' has been the presence there of shatter-cones, which are characteristic rock structures said to be produced exclusively by a violent shock, presumably an impact. Last time I went to Vredefort I photographed

Atmospheric meteors?

Aristotle, who lived from 384–322BC, believed meteors to be vapours which had been sent out from volcanoes and had accumulated as masses in the upper air. The masses were dragged around by the rotation of the sky, and were heated to incandescence; they became comets if slow-moving, meteors if quick-moving.

one particularly good specimen of a shatter-cone; you cannot mistake it.

I also went up in a helicopter to survey the Ring. The results were very striking. In places the wall is quite obvious; in others it has been almost levelled, so that it looks discontinuous. Still, from a height of thousands of feet you can trace its outline all the way round the circumference. Geologists have paid close attention to it for many years, and have made significant discoveries. First, the crater is linked with the 'local geology'. Secondly, the form is not characteristic of collision. All in all, the investigators are unanimous in claiming that the Ring is volcanic, and not an impact structure at all. The same is no doubt true of other alleged impact craters found in the official lists.

Next in our travels let us go to another part of the African Continent: Namibia. Close to Grootfontein there is the world's largest known meteorite. It is not an iron, but a stone; it has a weight of over 60 tons, and it is sitting just where it fell in prehistoric times. It has produced no crater at all, though we cannot be too sure about the effects of erosion over the ages; the Grootfontein area is much less durable than the desert of Arizona.

ENCKE'S COMET

Comets move round the Sun, most of them in very eccentric paths. Because they depend mainly upon reflected sunlight, they can be seen only when reasonably close to the Sun and the Earth — that is to say, in the inner part of the Solar System. The only periodical comet which can be followed throughout its orbit is Encke's, which has an interesting history. Whether or not it is associated with the Siberian missile of 1908 is debatable, but there is little doubt that it is the parent comet of the annual Taurid meteor shower. It has a period of 3.3 years, with its distance from the Sun ranging between approximately 32,000,000 miles and 380,000,000 miles. At its closest-in it is nearer to the Sun than the orbit of Mercury, while at its furthest it moves out into the asteroid zone; it never recedes as far as Jupiter, whose mean distance from the Sun is 483,000,000 miles.

It was originally found by the French astronomer Pierre Méchain, in January 1786, when it was a faint naked-eye object. The next observation came in November 1795, when it was seen by Caroline Herschel, the sister of William Herschel and herself an excellent comet-hunter. The comet was not recognised as being periodical, but it was seen once more in 1805, when it was discovered independently by three observers — Pons, Huth and Bouvard. Jean Louis Pons, one of the most famous of all comet discoverers, found it again in 1818, and this time the orbit was computed by J. F. Encke of the Berlin Observatory, who predicted a return for 1822. It was duly found, by C. L. Rümker, so near to Encke's calculated position that the comet has since been known by his name. Since then it has been seen at every return except that of 1944, when it was badly placed and when all observations were hampered by wartime conditions. The 1990 return was its fifty-fifth observed appearance since the original discovery by Méchain. Because of its small orbit, there is no real problem in photographing the comet even when it is at its furthest from the Sun, as was first achieved as long ago as 1913. There is no other comet with a period of less than five years.

Every time that a periodical comet passes through perihelion (its closest point to the Sun) it loses a certain amount of material by evaporation. By cosmical standards, therefore, a comet is short-lived, and most of the comets which come back regularly have become faint. Encke's is one of the brighter examples, since it can at times reach naked-eye visibility. A short tail may develop; in 1961 the tail-length amounted to two degrees of arc, which is approximately four times the apparent diameter of the Full Moon. In 1984 the maximum length of the tail was about one degree.

There have been suggestions that Encke's Comet is becoming steadily fainter, but this is by no means definite. It does not seem likely that it will 'die' before the end of the century, as was once proposed, and astronomers would be sad to see the last of it; after all, it is an old friend!

Shatter-cone at Vredefort. Shatter-cones, formed by tremendous pressure, are often said to be unique to impact craters, but this cannot be so, as this Vredefort example shows

Next, I must waft you temporarily to the tundra of Siberia, and ask you to look back in time to the morning of 30 June 1908. Suddenly a flaming object shot across the sky. It was brighter than the Sun, and it moved quickly; when it vanished, there was a tremendous explosion which was heard over hundreds of miles, and sent shock-waves all round the world. Something had hit the Tunguska area, and had caused widespread devastation.

For years nobody could decide what had really happened. Russia was not in the best of conditions at the time, and it was not until 1927 that an expedition led by Leonid Kulik managed to reach the area, which is extremely inaccessible even today. Kulik, naturally, expected to find a meteorite crater. He did not. Instead, he found an area in which large pine-trees had been blown flat. All the trees pointed away from the centre of the disturbance, and there was abundant evidence that there had been intense heat.

What was the answer? Kulik found no meteorite fragments; none has ever come to light. The general consensus of opinion is that the Tunguska missile was not a meteorite, but a piece of icy material —

Shower of stones

In 1802 a shower of 'stones' fell at L'Aigle, in France, and the French astronomer J. B. Biot was able to prove that they came from beyond the Earth. Not everyone was convinced. In 1806, for example, President Thomas Jefferson of the United States said: 'I could more easily believe that two Yankee professors would lie than that stones would fall from heaven.'

METEORITES

Meteorites are not associated with comets or with shooting-star meteors, and probably come from the asteroid belt. There may be no difference between a large meteorite and a small asteroid, though strictly speaking the term 'meteorite' is restricted to objects which have actually landed on the surface of the Earth.

Broadly speaking, meteorites are classified as stones (aerolites), stony-irons (siderolites) or irons (siderites), though there are many sub-divisions. When cut and etched with acid, an iron will show the characteristic 'Widmanstätten patterns', not found elsewhere. Over 80 per cent of the stones are chondrites, which contain spherical fragments of minerals (chondrules) with radiating structure; the remainder are achondrites, coarser-grained and with little free iron. Of special interest are the carbonaceous chondrites, which contain carbon compounds together with organic matter such as hydrocarbons.

It has been estimated that almost 20,000 meteorites weighing 3oz or more fall annually, but most of them come down in oceans or deserts and are lost to science. However, there are over 2,800 meteorites assembled in various collections, and more are being found regularly, Antarctica being a particularly profitable hunting-ground.

Major falls are rare, which is perhaps rather fortunate! All the known meteorites weighing more than 10 tons are irons, as follows:

	Weight (tons)
Hoba West, Grootfontein, Southern Africa	60
Ahnighito ('The Tent'), West Greenland	30.4
Bacuberito, Mexico	27
Mbosi, Southern Africa	26
Agpalik, Greenland	20.1
Armanty, Outer Mongolia	±20
Willamette, Oregon, USA	14
Chupaderos, Mexico	13
Campo del Cielo, Argentina	13
Mundrabilla, Western Australia	12
Morito, Mexico	11

Most of these have been collected, though the Hoba West meteorite is still lying where it fell in prehistoric times. The largest known aerolite fell in Manchuria in 1976, and weighs 1,766kg.

Nine well-authenticated meteorite craters over 200ft in diameter have been found. They are:

	Diameter (ft)
Meteor Crater, Arizona	4,150
Wolf Creek, Australia	2,800
Henbury, Australia	650x360
Boxhole, Australia	570
Odessa, Texas, USA	570
Waqar, Arabia	330
Oesel, Estonia	330
Campo del Cielo, Argentina	240
Dalgaranga, Australia	230

No doubt other large, less well-defined structures are also of impact origin.

either the head of a small comet, or else a fragment of the well-known periodical comet which we all call Encke's Comet, an old friend which comes back toward the Sun every 3⅓ years and has been known for over a century and a half.

Needless to say, the scientific fringe was quick to take the matter up, and there were all sorts of weird suggestions, ranging from a secret atom-bomb test to the results of an accident to a visiting spacecraft from another planet. Yet it does seem that the comet explanation is valid. No doubts at all are attached to the other major impact of our own century, that of 1947; again it was in Siberia, this time in the Vladivostok area, but the meteorite broke up before landing, producing many fragments and small craters.

It is just as well that these crater-forming impacts are rare. If the Siberian missile had hit a city, the death-toll would have been colossal. And if a mass of iron several hundred yards across came down in, say, the centre of Manchester, there would not be a great deal left of Birmingham. No doubt there have been major impacts in the

The Hoba West meteorite in Southern Africa; it weighs over 60 tons (Ludolf Mayer)

past, but I can assure you that there is no need for alarm; relative to the immensity of space, the Earth is a very small target.

I went back to Meteor Crater on my last visit to Arizona. Apart from improvements to the museum, it was exactly as before; after all, the Crater is ancient by any normal standards. It is a quiet spot, and a peaceful one; it is not in the least sinister (despite the Indian fears about it) and it has many moods. Seen under the heat of the mid-day sun, it is utterly different from its aspect under the Moon or the stars.

But things were very different once. Stretch your imagination and visualise what must have been the scene here so long ago — an occurrence which was seen by no human eye, since humans had not arrived here, but which might have been seen by some animals. Even then the area was infertile, but suddenly, out of the sky, there must have appeared a blinding light, far outshining the Sun and marking one of the most dramatic episodes of the near-modern world. In a flash it must have been down; a scorching, searing burst of heat, an earth-shattering explosion, and then the shuddering and quaking as

Siberian débris. Pine-trees blown flat by the Siberian impact of 1908, which is now thought to have been icy rather than meteoritic

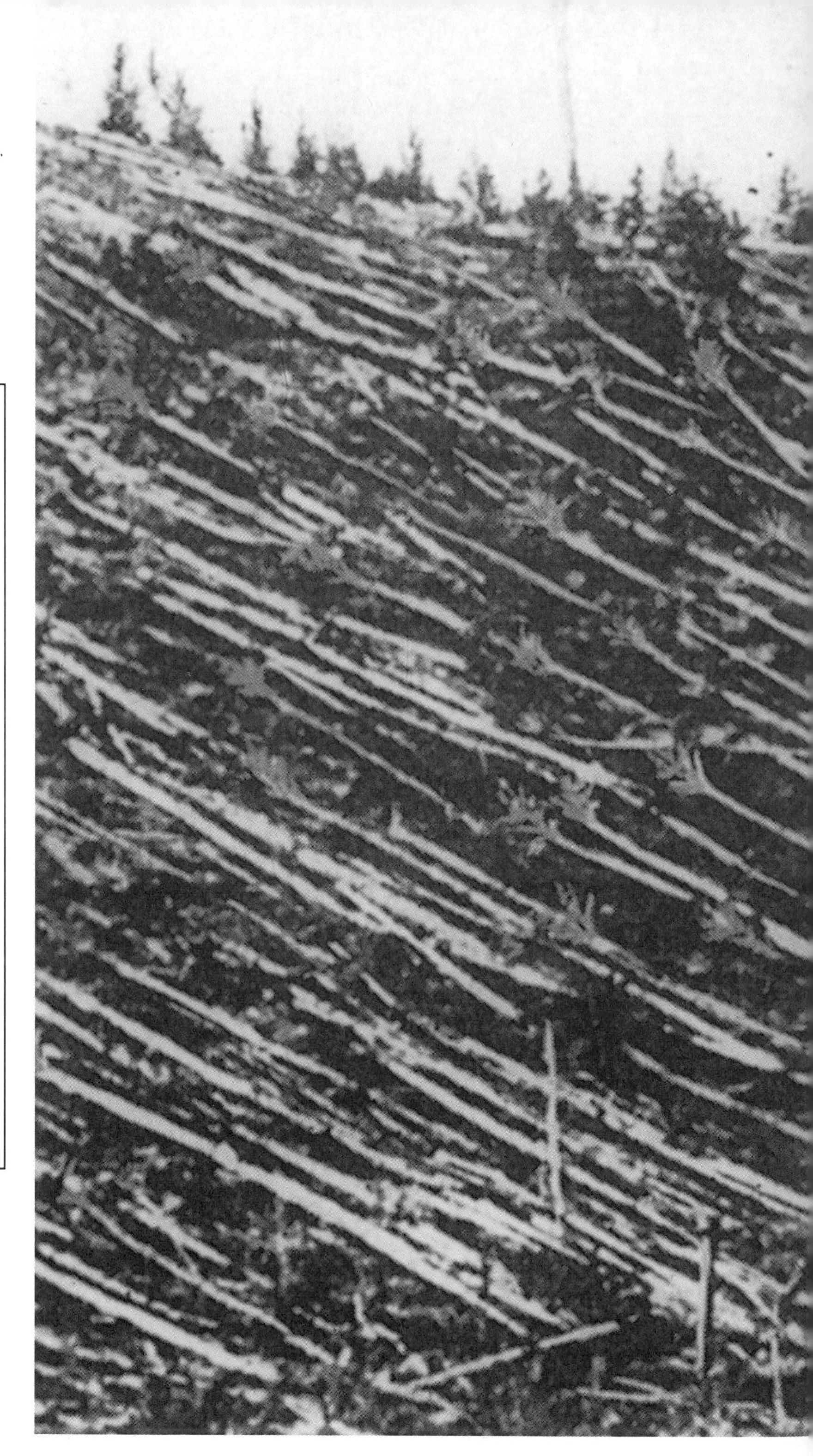

Flying saucer?

The flying saucer explanation of the Tunguska missile seems to have originated in 1959 in an article by a Russian science-fiction author, Alexander Kazantsev, who claimed that the object was a visiting spacecraft which made an accidental landing and showered nuclear débris in all directions. 'At the moment of disaster,' wrote Kazantsev, 'the temperatures rose to millions of degrees. Elements were vaporised, and in part carried into the upper atmosphere, where they caused luminescence and then fell to the ground, with radioactive effects.' Subsequently there were stories about mutations in the local plant and even animal life, which, predictably, have never been confirmed. All of which goes to show that events of this kind attract 'UFOlogists' and their kind in the same way that wasps will collect round a jamjar.

the ground heaved and twisted. If there had been any life for miles around, it could not have survived. In less than ten seconds, that part of Arizona was transformed from a placid landscape into an inferno.

How long did it take for the effects to die away? We do not know, but the Earth is resilient, and recovers amazingly quickly. Within tens of years, or even less, what little scrub had existed pre-Crater must have returned, and the tremors long since died away. Since then the

Crater has been undisturbed apart from the mercifully brief attempts to mine it and the much more welcome stream of present-day visitors coming to marvel at it.

Yes, I feel that Svante Arrhenius may well have been right when he called it 'the most interesting place on earth'. When you see it now, it is hard to realise that for a few moments in the remote past it was also the most terrifying place in the whole of the world.

Rocks and rubble

A meteorite is not always easy to identify at first sight, as I found to my cost when presenting a television programme about them. People began sending me 'meteorites' of all kinds, and I accumulated a vast collection of stones, rocks, rubble and the like, including an object which proved to be a very ancient Bath bun. Not one of the samples was truly meteoritic.

I also remember making a journey to the area of Stornoway, in the Hebrides, to investigate a report that a meteorite had fallen there and had blown water from one small loch uphill into another. I was suspicious from the outset, because the original report claimed the 'meteorites had fallen there before', and the chances of two impacts close together are very low indeed. I was right; the object was an 'erratic' — that is to say, a boulder which had been carried far away from its place of origin, probably by a glacier — and the water from one loch to another had been downhill, not up, so that it could be attributed to a minor gaseous outbreak in the boggy ground.

Meteorite casualty

Only one creature has definitely been identified as having been killed by a meteorite: this was an Egyptian dog in 1912. However, there have been a few narrow escapes. On 30 November 1954, Mrs E.H. Hodges, of Sylancanga, Alaska, was slightly injured in her arm when a 9lb meteorite crashed through the roof of her bedroom.

THE ZODIACAL LIGHT

An interesting but rather elusive phenomenon of the night sky is the Zodiacal Light, a dim, cone-shaped glow rising from the horizon soon after sunset or soon before sunrise. From Britain it is best seen during evenings in March or mornings in September, and is never conspicuous, though from countries with clearer skies it can sometimes become almost as bright as the Milky Way. It has been known since 1683, when it was first reported by the Italian astronomer G.D. Cassini.

The Zodiacal Light is caused by sunlight illuminating the very thinly spread 'dust' which extends through the Solar System, chiefly in the main plane. A still fainter extension, stretching from the cone across the sky, is called the Zodiacal Band. Finally there is the Gegenschein or Counterglow, which lies exactly opposite to the Sun in the sky and takes the form of an extremely dim patch, about the same apparent size as the full moon. From Britain I have seen it only once, during World War II, when the whole country was blacked out as a precaution against German air-raids, and the sky was pleasingly dark. I have never seen a really convincing photograph of it.

The Zodiacal Light taken from Los Muchachos, Canary Isles, in 1990

Moonshine

Plundered telescopes

Johann Hieronymus Schröter was the first really great observer of the Moon. He was not a professional astronomer, but was chief magistrate of the little town of Lilienthal, near Bremen. Unfortunately his observatory was destroyed in 1814 when the French invaded Germany, and even his telescopes were plundered; they had brass tubes, and the soldiers mistook them for gold.

Go out on a clear night, under a full moon, and you may think that you can see almost as well as you can by day. It comes as rather a surprise to find that it would take about half a million full moons to equal the light of the Sun. Moreover, moonlight is second-hand sunlight; the lunar rocks are grey, and reflect less than ten per cent of the solar rays. Yet the Moon is of true importance to us, because it is our faithful companion, and stays with us as we travel round the Sun. We cannot picture an Earth without its Moon.

At the moment it is the only world to have been reached by Man. We have found out what it is like; we have brought home samples of its rocks, and analysed them in our laboratories and we have looked for signs of life. We have not found them, and rather to our disappointment we have been forced to realise that the Moon is a world which has never known living creatures. This does not make it any the less intriguing, but it has led to a change in our ideas, because not so many centuries ago it was widely believed that 'Moon men' were real.

Even with the naked eye it is easy to see the main dark patches on the Moon, and in past times it was natural to think that these dark regions were seas. When telescopes were turned moonward, in the early part of the seventeenth century, the dark regions were given romantic names such as the Sea of Tranquillity, the Bay of Rainbows and the Ocean of Storms; the names are still used, even though we have long since found out that the seas are nothing more than bone-dry lava-plains. Going back still further, it was often thought that the Moon might be capable of supporting intelligent beings. Plutarch, the Roman writer who lived between AD46 and 120, wrote a book with the title (in English) *The Face in the Orb of the Moon,* in which he claimed that the Moon was 'terrestrial, and inhabited like the Earth, peopled with the greatest living creatures and fairest plants'. However, the first account of a lunar voyage was due to a Greek satirist, Lucian of Samosata, who was born about the same time as Plutarch died.

Lucian's book was called the *True History,* but there was nothing true about it; Lucian himself was the first to point out that it was made up of nothing but lies from beginning to end. In his story, a party of

Tsiolkovskii, from Orbiter 3. This great formation, on the Moon's far side, seems to be intermediate in type between a crater and a 'sea'

sailors passing through the Pillars of Hercules (known to us as the Straits of Gibraltar) were caught in a waterspout, and hurled upward so violently that they went on ascending for seven days and seven nights, finally landing on the Moon. They were promptly met by men riding on large, three-headed griffins, and were taken to the King of the Moon, where they learned that they had arrived at a critical moment; the King of the Moon was about to go to war against the King of the Sun over a dispute as to who should have first rights on Venus (the planet, I hasten to add, not the goddess). The lunar army included such unusual recruits as cabbage-fowls, which were vast birds with cabbage-leaves instead of wings; riders mounted on fleas, each of which was as large as a dozen elephants; sparrow-fighters, crane-riders and so on. Unfortunately, the Sun King's cloud-centaurs were more than a match for the opposition, and the war came to an abrupt end.

Lucian was nothing if not imaginative. 'All Moon men live on the same kind of food. They light a fire and roast frogs, of which they have large numbers flying about in the air . . . Any Moon Man who is bald and hairless is regarded as handsome. Long-haired people are abhorrent to them. There are no women on the Moon; when a Moon man grows old he does not die, but dissolves in smoke.'

All this is great fun, but it is not science, and was never intended

Antics

According to an account which seems to be authentic, an eighteenth-century amateur reported that with the aid of his telescope he had seen large animals walking about on the surface of the Moon. They turned out to be ants in his eyepiece.

Lunar names

The entire lunar scene is dominated by craters. In 1651 the Italian astronomer Riccioli drew a map of the surface, and gave the craters names of important people, usually astronomers — a system which we still use, though naturally it has been extended since Riccioli's time, and inevitably astronomers of later years came off second best (Newton, for example, has been allotted a deep but very badly placed formation in the southern uplands). Some of the names sound rather unusual. You would hardly expect to find Barrow, Birmingham, Billy and Birt, but they are all there, together with Julius Cæsar, Atlas and Hercules. Isaac Barrow was the Cambridge professor who resigned his Chair in favour of Newton; John Birmingham was an Irish amateur astronomer of the nineteenth century who discovered an interesting new star and was also noted for his studies of the Moon; Jacques de Billy was a seventeenth-century professor of mathematics, and William Radcliffe Birt, who lived from 1804 to 1881, was a well-known lunar mapper. Julius Cæsar is there not because he ruled Rome,

continued opposite

to be. The same is true about the various legends about the Man in the Moon. I admit to having never been able to see the Old Man as clearly as most people seem to, though the arrangements of light and dark patches can be said to conjure up a vague impression of a human form. Certainly there are legends in plenty. According to a German tale, the Old Man was a villager who was banished to the Moon when he was caught stealing cabbages; in Turkey, the Moon was a young bachelor who was engaged to the Sun and shone in the daytime, but the girl-Sun was afraid of the dark, and persuaded her fiancé to change rôles. A myth from Greenland tells how the Sun and the Moon were brother and sister, and for some reason or other the Sun rubbed soot in his sister's face, producing the dark patches we now see. Rather naturally, the Moon chased him, but she can never catch him, because she cannot fly so high. Every few weeks she feels the need for a rest, so she comes back to the ground, climbs into a sleigh drawn by four dogs, and goes seal-hunting. After breakfasting off several seals she regains her strength, and the Moon once more appears in the evening sky to start the chase all over again.

Moon-worship was common, and in parts of Central Africa it

The lunar scene: Archimedes to Julius Cæsar. Julius Cæsar is the distorted formation above centre, slightly to the left; above it, to the right, is the Hyginus Rill; below, the border of the Mare Serenitatis

Eclipse of the Moon. The Moon is indeed 'losing its light'! (Commander H. R. Hatfield)

lingers on even today. But serious ideas about possible life come into a different category. There were some writers, such as the Swedish mystic Emanuel Swedenborg, who believed as a matter of principle that all worlds must be inhabited, but on a more sober note we must probably begin in the early seventeenth century, and consider the views of Johannes Kepler.

Kepler was a curious mixture. It has been said that he had one foot planted firmly in the past, and the other equally firmly in what we may term the near-modern age. He was a practising astrologer (though whether he really believed in it is another matter), and his first book, the shortened title of which is *Mysterium Cosmographicum,* contains some ideas which sound strange today; for example, he linked the distances and movements of the planets with the five regular solids of geometry in a way which was none too easy to credit even in his own time. But he was also a brilliant mathematician, and it was he who first realised that instead of moving round the Sun in circular paths, the planets — including the Earth — move in elliptical orbits. His Laws of Planetary Motion, the first two of which were published in 1609 and the third in 1618, really paved the way for Newton.

Kepler, then, was a true scientist. He also produced a book which could be classed as science fiction, but was really intended to be genuine science with a thin disguise. It was published posthumously in 1632, two years after Kepler had died — rather sadly; he was

continued

but because he was responsible for a major revision of the calendar, while Atlas and Hercules are a couple of Olympians.

Personal considerations could not be ruled out. Riccioli took care to name large and prominent walled plains after himself and his pupil Grimaldi, but he was no believer in the Copernican theory that the Earth moves round the Sun, and so he 'flung Copernicus into the Ocean of Storms', though admittedly he gave the name to one of the most majestic formations on the Moon.

Lunar poetry

During the 1940s it is said that a housemaid in the service of a well-known poet penned the following immortal lines:

O Moon, lovely Moon with
the beautiful face,
Careering through the
bound'ries of space,
Whenever I see you, I think in
my mind
Shall I ever, O ever, behold
thy behind?

Whether she lived to see the first Luna 3 pictures does not seem to be known.

always desperately short of money, and he had family troubles as well. All in all, his life was not a happy one. His story, the *Somnium* or Dream, describes a journey to the Moon undertaken by the intrepid Icelandic hero Duracotus. The mode of travel would hardly appeal to NASA, because Duracotus reaches the Moon by demon power; during a lunar eclipse he is carried across on the bridge formed by the Earth's shadow. Yet the description of the Moon itself fitted the facts as far as Kepler knew them, and it contains some sound seventeenth-century science.

In particular, Kepler knew that the Moon always keeps the same face turned toward the Earth, because its revolution period is the same as its axial rotation period: 27.3 Earth-days. Accordingly, Kepler divided his Moon into two zones, Subvolva and Privolva. From Subvolva, the Volva (Earth) is always visible; from Privolva, never. The Moon is a world with violent changes of temperature, towering

THE OTHER SIDE OF THE MOON

The Moon takes 27.3 days to complete one orbit; it also spins on its axis in 27.3 days. At first sight this might seem to be a remarkable coincidence, but in fact it is nothing of the kind. The rotation of the Moon is captured, or synchronous, because of tidal friction over the ages.

In the early history of the Solar System, the Earth and the Moon were closer together than they are now. The Moon was not rigid, as it is today; it was spinning round much more rapidly than it does now, but the Earth raised strong tides in it, so that as the Moon rotated there was a tendency for a 'tidal bulge' to be produced in the direction of the Earth. It was a situation rather like that of a cycle wheel spinning between two opposite brake-shoes. The Moon's rotation was slowed down, and eventually stopped altogether with respect to the Earth, so that the same hemisphere was turned toward us all the time.

To show what is meant, walk round a chair, turning as you go so as to keep your head turned 'chairward'. After one full circuit you will have faced every wall of the room, but anyone sitting on the chair will never have seen the back of your neck. Similarly, we on Earth never see the 'back' of the Moon.

From this, it might be thought that only 50 per cent of the total surface would be available for our inspection. Actually we can examine a grand total of 59 per cent, though, of course, never more than 50 per cent at any one moment. This is because of effects known as librations.

Although the Moon rotates on its axis at a constant rate, it does not move in its orbit at a constant rate; following Kepler's Laws, it moves fastest when closest to us — and its orbit is appreciably eccentric, with the distance ranging between 221,460 miles ('perigee') and 252,700 miles ('apogee'). Therefore, the amount of rotation and the position in orbit become periodically out of step, and the Moon seems to oscillate slightly; first a little of the eastern limb is exposed, then a little of the western limb. This is *libration in longitude.* There is also a *libration in latitude,* because the Moon's equator is tilted to the plane of its orbit by 6°, and finally there is a *diurnal libration,* because we are observing from the Earth's surface rather than its centre. The sum total of all these librations is that only 41 per cent of the whole surface is permanently out of view, though admittedly the edges of the disk — the so-called libration zones — are so foreshortened that they are difficult to map.

This was the situation until October 1959, when the Russians launched their unmanned spacecraft Luna 3. Pictures were obtained of the hitherto-unknown far side, and were sent back by television techniques, confirming that the far side regions are essentially the same as the familiar areas — though there are no 'seas' on the scale of the Mare Imbrium or the Oceanus Procellarum, and there are subtle differences in colour and in crater distribution. Since then much better pictures have been obtained, notably by the unmanned Orbiters of 1966-7 and by the Apollo astronauts, so that we now have a detailed knowledge of the whole of the surface.

Note, incidentally, that although the Moon keeps the same face turned toward the Earth, it does not keep the same face turned toward the Sun, so that day and night conditions are the same everywhere on the Moon — apart from the fact that from the far side, the Earth can never be seen.

The crescent Moon (south is at the top). The small circular sea to the lower left is the Mare Crisium (Commander H. R. Hatfield)

mountains and deep valleys, which is of course quite correct. 'The hollows of the Moon, first seen by Galileo, are portions below the general level, like our oceans, but their appearance makes me judge that they are swamps for the greater part.'

The Moon-folk are of various types. Some are serpent-like, while others have numbers of fins to help in swimming, and others keep close to the ground. Most have fur; if a Moon creature is unwise enough to be caught in the open near midday its outer fur is singed, and the creature drops down as though dead, though at nightfall it revives and the burned fur simply drops away. 'The Privovans have no settled homes. During a single day they may wander, in large herds, right across their hemisphere. They use their legs, which are longer than those of camels, or flutter about with their wings, or sail in their boats upon the waters of the floods, or creep into the caves for protection. Some of them, with fishlike bodies, find respite from the heat of the Sun in the cool waters of the oceans and rivers. Most of the people of the Moon are divers, and all breathe very slowly, so that they are able to spend a long time under the surface of the water . . . All over the Moon lie great masses of acorns. The scales of these cones are burned by the Sun during the day, but in the evening the cones open, and new Moon-dwellers are born.'

It is difficult to judge how much of this Kepler really believed, but undoubtedly he did think that the Moon was inhabited. Four years after the *Somnium* came another Moon-voyage story, the *Man in the*

Approaching the Moon. Photograph taken from Eagle, the Command Module of Apollo 11, just before landing at Tranquillity Base (Neil Armstrong)

Lunar theories

Though opinion is still divided as to whether the lunar craters are due to volcanic action, meteoritic bombardment or (much more probably) a combination of both, other less plausible theories have been put forward from time to time. For example, the Austrian engineer Josef Weisberger solved the whole problem very neatly, simply by denying that there were any lunar mountains or craters at all; he claimed that they were merely storms and cyclones in a dense atmosphere. I never met Herr Weisberger, who died just after World War II, but I gather that it was rather difficult to argue with him, because anyone who disagreed with him was automatically classed as a mortal enemy.

Another Austrian, Hans Hörbiger, was convinced that practically everything in the universe was made of ice, so that even the stars were ice-blocks in the sky rather than suns. Our ice Moon was not our first attendant; there had been at least four previous Moons, which had crashed onto the Earth's surface and caused great damage. (Needless to say, the last of these collisions caused the Biblical Flood.) Hörbiger's 'universal ice' theory was widely supported in pre-war Germany, and at one stage the Government had to issue a statement pointing out that one could still be a good Nazi without supporting it.

Coral reefs were also suggested, but the king of all lunar theories was unquestionably yet another engineer, Sixto Ocampo of

continued on page 52

Moone, by an English bishop named Francis Godwin. It became very popular, and is still worth reading. There is little science in it, since his hero is carried moonward on a raft towed by wild geese, and his picture of the lunar inhabitants is as light-hearted as Lucian's had been so long before, but there is one particularly lovely touch in it. The Moon-men are gigantic, towering to at least 30ft in some cases, but they are also benign; Moon children who fall short of the high moral standards required are promptly exiled to Earth, where there is already so much depravity that a little more cannot possibly matter. The Moon-men are able to tell at once whether or not a newly-born child is likely to prove acceptable. If it shows signs of latent wickedness, it is dispatched without delay:'The ordinary vent for them is a certain high hill in the North of America, whose people I can easily believe to be wholly descended from them.' Evidently the good Bishop had a poor opinion of the people of the New World.

At about the same time came a different account by another bishop, John Wilkins. This was *Discovery of a New World,* made up of two Discourses, one 'Tending to prove it is Probable that there may be another Habitable World in the Moon'. Wilkins was perfectly serious, and he was also a patriot, so that he tried to persuade the Government to conquer the Moon for the benefit of the English nation. The Moon had, he said, 'many excellent uses', with mountains to 'tame the violence of great rivers, and break the force of the seas' inundation, for the safety of the inhabitants, whether beasts or men'. He conceded that the Moon people might not be human in form, 'but some other kind of creatures which bear some proportion and likeness to our natures, or it may be that they are of quite different nature from anything here below, such as no imagination can describe'.

Wilkins believed that the Moon had breathable air, and in his time there was no reason to think otherwise, but from around the start of the nineteenth century we come to a period when true astronomy began to take over. Telescopes had improved, and the first reasonably accurate lunar maps had been produced.

Not that there was any lack of wild ideas — and, surprisingly, the oddest of all came from the man who is generally regarded as the

On the Moon with Apollo 11. Colonel Edwin Aldrin sets up a seismic experiment at Tranquillity Base

continued from page 50
Spain. This time the craters were produced during a nuclear war between two races of Moon-men, who wiped each other out in the process. The fact that some craters have central peaks while other have not shows, of course, that the two sides used different kinds of bombs. The last real bangs on the Moon 'fired' the lunar seas, which fell back to Earth *en bloc* and caused — yes! — the Biblical Flood.

Señor Ocampo submitted his work to the Barcelona Academy of Arts and Sciences, but they rejected it. Ocampo then published it himself, with a cross little Foreword to the effect that an unscrupulous British astronomer had stolen his theory and was about to claim it as his own, thereby depriving Spain of the glory of the discovery. His life's work was done, and he died almost immediately.

greatest astronomical observer of all time, William Herschel. In 1781 Herschel, then an unknown amateur, discovered the planet Uranus, thereby doubling the size of the known Solar System. Later he discovered thousands of double stars, clusters and nebulæ; he was the first man to work out a fairly good idea of the shape of the Galaxy; he made the largest and best telescopes of his time, and during his long life he received every honour that the scientific world could bestow. Yet he believed that the habitability of the Moon was 'an absolute certainty', and he went so far as to claim that there were intelligent beings living in a cool, calm region below the surface of the Sun. Just how he can have believed this defies logic, and few of his contemporaries agreed with him, so that references to the Sun-men were usually edited out of his scientific papers before they appeared in print.

However, the Moon was a different proposition, and in Herschel's time there was still no reason to deny the possibility of lunar life of some sort. Everything really hinged upon whether or not there was any substantial atmosphere. One man who believed so was Johann Hieronymus Schröter, a German contemporary of Herschel's who was the first really great observer of the Moon (and who, incidentally, often used a telescope which Herschel had made for him). According

THE APOLLO MISSIONS

Sending an expedition to the Moon is not so straightforward as might be thought. Only the rocket can be used for space-travel, because it alone can function in 'empty space', but to send a single vehicle direct from the Earth to the Moon, land it there and then bring it back would be impossible by our present techniques.

An Apollo craft carries three astronauts. Following the launch, by a very powerful compound rocket, the vehicle is put into a closed path round the Moon, after which two of the crew members make the final descent in the 'Lunar Module' which has been brought with them. When the time comes to depart, the lower stage of the Module is used as a launching-pad; the upper section uses its single motor to blast back into orbit, where it rejoins the main part of the space-craft ready for the homeward journey. The main weakness of this procedure is that there is no provision for rescue. The ascent engine of the Lunar Module has to work perfectly; there can be no second chance.

There were two full-scale rehearsals. In December 1968 Apollo 8 made a 'round trip', and in the following May 1969 the Lunar Module of Apollo 10 was tested in the neighbourhood of the Moon. All went well, and the first landing was achieved in July 1969 when first Neil Armstrong, then Edwin Aldrin stepped out on to the bleak rocks of the lunar Mare Tranquillitatis. Nobody will ever forget Neil Armstrong's words: 'That's one small step for a man — one giant leap for mankind', and nobody has ever bettered Edwin Aldrin's description of the lunar scene as 'magnificent desolation'.

Armstrong and Aldrin carried out a preliminary 'Moon walk', lasting for just over two hours; their first task was to collect rock samples, after which they set up scientific instruments which were designed to continue sending back data after the astronauts had left. Despite their cumbersome-looking pressure-suits, without which they could not have survived for an instant, they had no trouble in moving around. Since they had only one-sixth of their normal Earth weight, they gave the impression of walking about in slow motion. As had been expected, there were no communication problems, and television viewers all over the world were able to see and hear Armstrong and Aldrin as they walked around on the lunar surface.

As expected, the samples of lunar material brought back were essentially similar to the volcanic rocks of Earth, and there were no major surprises. There was no sign of life, either past or present, but as a precaution the astronauts were strictly quarantined until a full examination had been carried out. The chances of

to Schröter, the lunar atmosphere was thick enough to support clouds.

Then, in 1835, came a truly hilarious episode which is still remembered. It was a hoax, but a very skilful one. The victim was William Herschel's son, Sir John, who had gone to the Cape of Good Hope to carry out a systematic survey of the far-southern stars, which can never be seen from Europe and which had therefore been rather neglected. He sailed from England in the early part of 1834, and remained at the Cape until 1838, by which time his work had been well done. He did not mean to pay any particular attention to the Moon or planets, which can be seen just as well from the northern hemisphere as from South Africa, but Richard Locke, an enterprising reporter on the New York *Sun,* had a bright idea. Herschel was on the other side of the world; communications were slow; who was there to check any claim which might be made?

Locke took his chance. On 25 August 1835, the *Sun* came out with a startling headline: 'Great Astronomical Discoveries Lately made by Sir John Herschel at the Cape of Good Hope'. Apparently Herschel had invented a new type of telescope which was powerful enough to show the Moon in amazing detail. Locke was a clever journalist, and his article sounded remarkably plausible. It was well known, he wrote, that the chief limitation of any telescope is that it cannot collect

False alarm

Soon after the launch of Sputnik 1, an agitated American householder rang the Pentagon to say that a satellite had landed in a tall tree in her garden. Security officials were sent to investigate. The object tuned out to be a large balloon, with 'upski' printed on the top and 'downski' on the bottom.

bringing home anything harmful were very slight, but they were not nil (remember Professor Quatermass!) and not until a further successful mission was quarantining abandoned as unnecessary.

Apollo 12 followed in November 1969; Astronauts Conrad and Bean even went over to an old unmanned space-probe, Surveyor 3, which had been on the Moon since 1967, and brought back parts of it. This time there were two Moon walks, and further samples were collected.

Apollo 13, of April 1970, was a near-disaster. On the outward journey there was an explosion in the main spacecraft; the lunar landing was abandoned, and it was only by a combination of courage, skill and luck that the astronauts returned unharmed. It was a timely reminder that space travel is anything but safe; there had been a tendency for some people to adopt a rather blasé attitude.

All the last four Apollo missions were successful. No 14 (February 1971) was commanded by Alan Shepard, who had been the first American in space only ten years earlier. He and his companion, Edward Mitchell, took a 'cart' with them to help in carrying equipment and setting up experiments, so that they were able to cover a distance of over a mile in the area of the ruined crater Fra Mauro.

With Apollo 15 there was a new departure; the astronauts took an electrically driven 'Moon car', so that they were able to drive across the lunar surface. David Scott and James Irwin landed in the foothills of the Apennine Mountains, and went to the very edge of one of the great lunar valleys. Nine months later, in April 1972, astronauts Young and Duke, in Apollo 16, explored the highlands of the Descartes region in the southern part of the Moon, and remained outside their Lunar Module for a total of over twenty hours.

The climax came in December 1972, with Apollo 17. The commander of the mission was Eugene Cernan; his companion, Dr Harrison Schmitt, was a professional geologist who had been trained as an astronaut specially for the mission. When they finally departed, having accomplished everything that they had hoped to do, the first phase in Man's direct exploration of the Moon was over.

Much had been learned. Now, less than twenty years later, plans are being made to set up a full-scale Lunar Base which will be of immense value; the Moon is an ideal site for a scientific base, and there are strong hopes that the next missions will be international. There is no reason why the Base should not be operational by the end of the century.

Because the Moon has no atmosphere and no 'weather', the equipment left on the surface by the earlier missions will remain intact — and no doubt some future astronauts will go up to the abandoned Moon cars, put in new batteries, and drive them away to be displayed in a lunar museum. It is an intriguing thought!

The far side of the Moon

While studying the Moon's movements, the nineteenth-century Danish astronomer Peter Andreas Hansen found some tiny irregularities from which he concluded that the moon was lop-sided, with the centre of mass 33 miles away from the centre of the globe. Consequently, all the air and water had been drawn round to the far side, which might well be inhabited.

Astronomers in general were unimpressed, and were even more sceptical about the claim by George Adamski, in 1953, that he had been round the Moon in a flying saucer and had seen little furry animals running about among the far-side craters. In a paper of my own, published in the same year (1953) well before the flight of Luna 3, I wrote: 'All the evidence we can muster shows us that the hidden side of the Moon is much the same as the side we can see. Seas may be lacking, but there must be mountains and craters in plenty, along with ridges, clefts and systems of bright rays.' I can modestly claim that I was right.

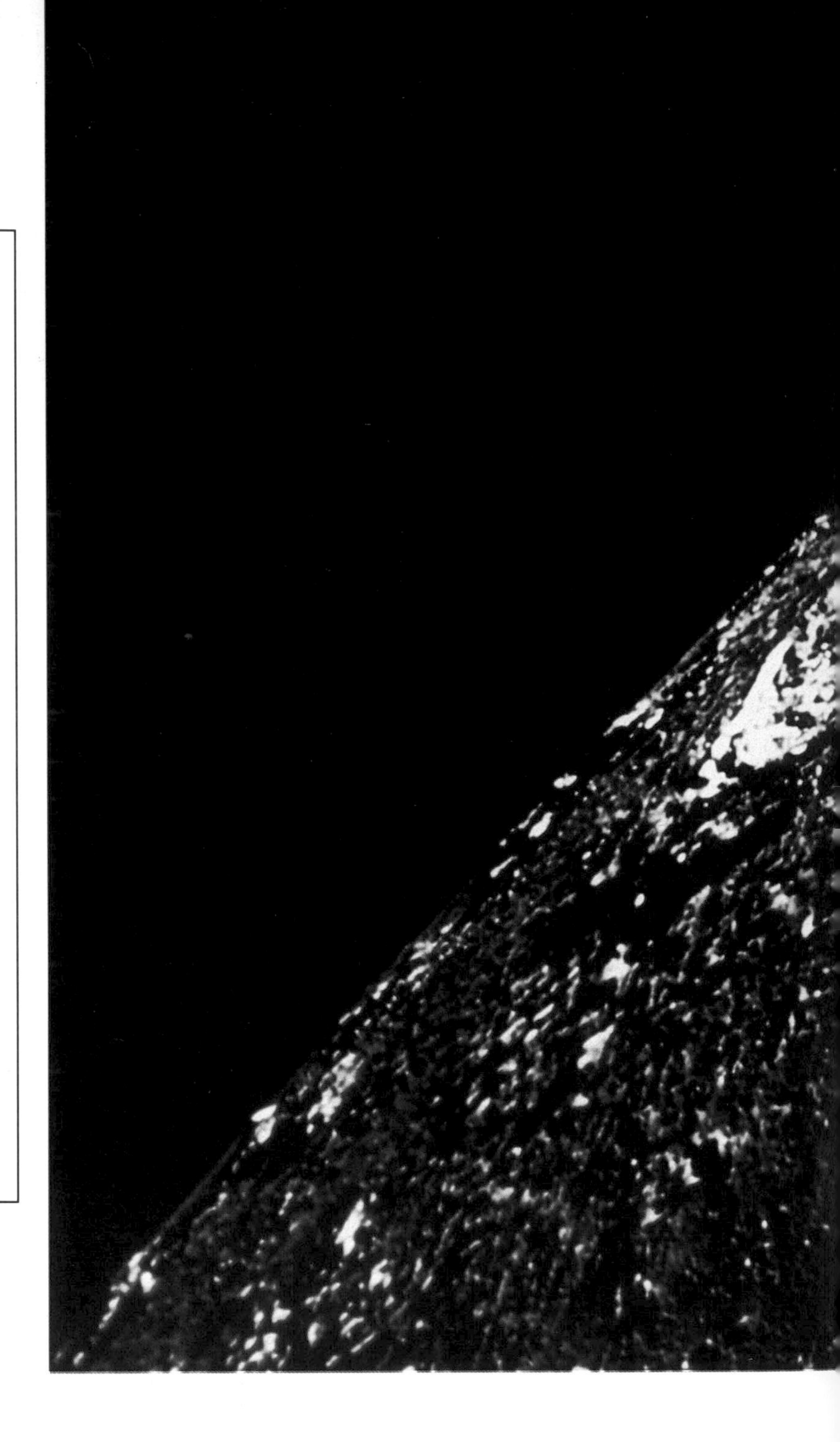

enough light for extreme magnification, but Herschel had overcome this by 'transfusion of artificial light through the foculpoint of vision'. In other words, you use the telescope to form an image, and then reinforce the image by a light-source in the observatory itself!

For the next six days the *Sun* kept up the good work. The lunar scenery was nothing if not varied. There was, for example, 'a lofty chain of obelisk-shaped or very slender pyramids, each composed of thirty to forty spires, every one of which was perfectly square'. Next came animals, of which one was 'of a bluish lead colour about the size

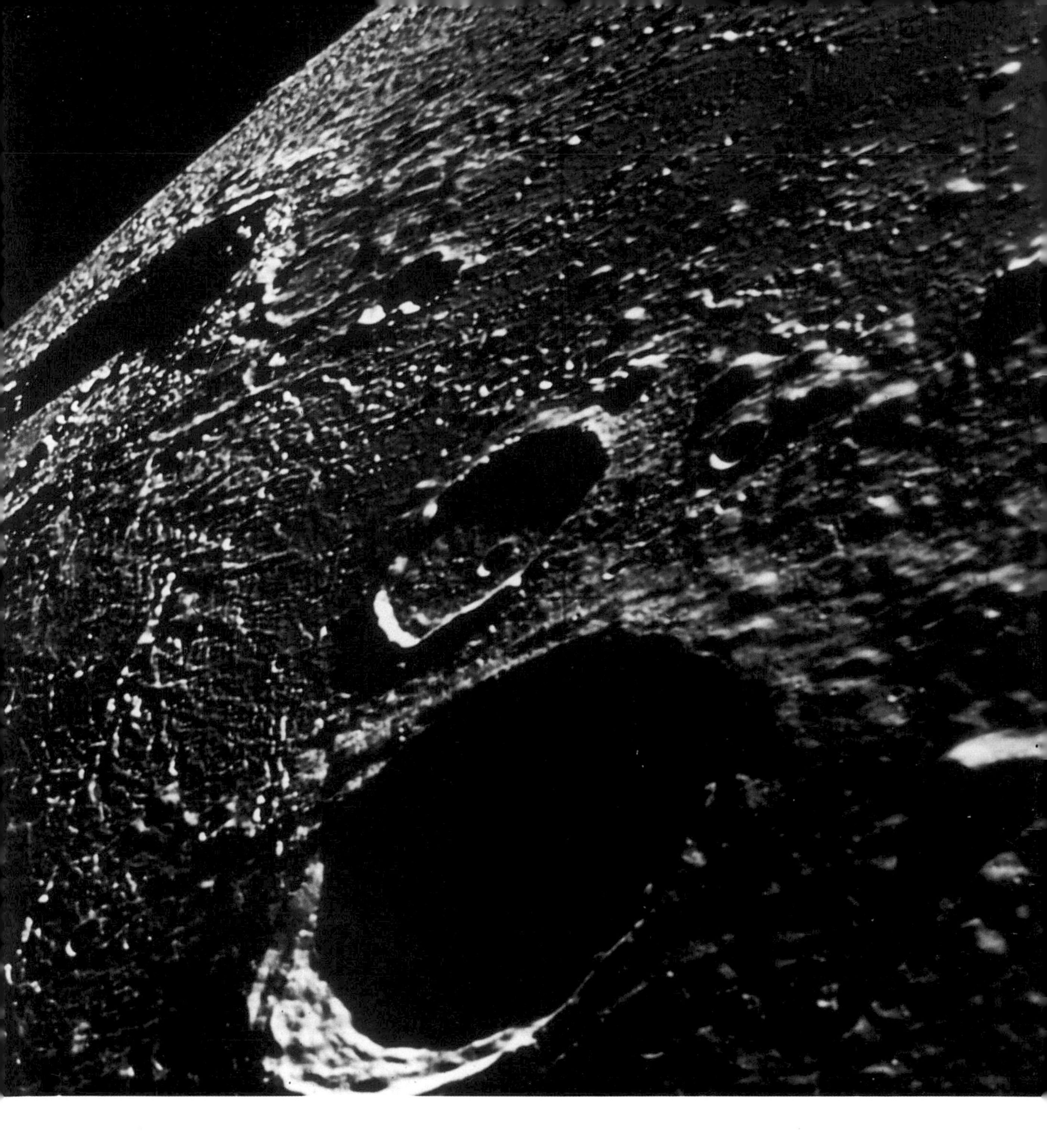

of a goat, with a head and beard like him, and a single horn . . . In elegance of symmetry it rivalled the antelope, and like him it seemed an agile sprightly creature, running at great speed and springing up from the green turf with all the unaccountable antics of a young lamb or kitten. This beautiful creature afforded us the most exquisite enjoyment.' Then there was 'a strange amphibious creature of spherical form, which rolled with great velocity across the pebbly beach'. Many exotic trees were catalogued, and in the crater Cleomedes there was 'a large white bird resembling a stork', while there were

Craters Copernicus and Reinhold from Apollo 12. (Copernicus is in the background)

The descending module of Apollo 14, with the Earth in the sky. The two astronauts were Shepard and Mitchell

Rocket pioneers

Rocketry has produced its full quota of colourful characters. One of these was a Chinese official, Wan-Hu, who is said to have made an extraordinary craft out of two kites, a saddle and forty-seven rockets, the rockets being fixed to a bamboo framework attached to the saddle. When all was ready, Wan-Hu called all his friends and relations to watch his ascent, and ordered his servants to fire all forty-seven rockets at once. They did — and nothing more was ever seen of Wan-Hu. The date is given variously as 1500, 1600 and 1603. If the story is true, Wan-Hu has the honour of being the first victim of experimental rocket research.

Another Russian rocket designer was named Kibaltchitch. In 1881 he produced detailed plans for a sort of platform with a hole in the centre, above which was a cylinder filled with gunpowder. His idea was to raise the cylinder vertically, and then control its movements by tilting the cylinder and altering the direction of thrust. Unfortunately for him, he was caught and (correctly) convicted of making the bomb used to assassinate the Czar of Russia, and he was given no chance to experiment further.

also 'hills pinnacled with tall quartz crystals, of so rich a yellow and orange hue that we first supposed them to be pointed flames of fire'. Aristarchus, the brightest feature on the Moon, was 'a volcanic crater, awfully rivalling our Mounts Etna and Vesuvius in the most terrible epochs of their reign . . . we could easily mark its illumination of the water over a circuit of sixty miles'.

The climax came on 28 August, with Locke's priceless description of lunar bat-men:

> Certainly they were like human beings . . . They averaged four feet in height, were covered, except on the face, with short and glossy copper-coloured hair, lying snugly upon their backs. The face, which was of a yellowish flesh-colour, was a slight improvement upon that of the large orang-utan . . . .The mouth, however, was very prominent, although somewhat relieved by the thick beard upon the lower jaw. These creatures were evidently engaged in conversation; their gesticulation, particularly the varied action of their hands and arms, appeared impassioned and emphatic. We hence inferred that they were rational beings.

Locke was clever enough to bring in 'science' now and then. Sometimes a higher-power eyepiece had to be used, sometimes it was necessary to turn up the hydro-oxygen burners to light up the faint image by 'artificial transfusion', and so on. Finally, the observations were brought to an end when Herschel forgot to cover up the main lens, so that when the Sun shone on it the lens acted as a large burning-glass and set fire to the observatory.

Nobody knew quite what to make of all this, but some eminent critics swallowed the bait hook, line and sinker. The New York *Times* declared that 'these new discoveries are both probable and plausible', while the *New Yorker* felt that the observations 'had created a new era in astronomy and science generally', which would certainly have been true. A women's club in Massachusetts wrote to Herschel asking him how they could be put in touch with the bat-men in order to convert them to Christianity. Others were more cautious. The hoax was soon exposed, and the *Sun* itself confessed on 16 September, but lingering doubts remained, and not for some months was the whole absurd business killed. Herschel took it all very well; fortunately he had a strong sense of humour.

Meanwhile, two Berliners, Wilhelm Beer and Johann Heinrich von Mädler, were working on a new map of the Moon. It was published in 1838, and was a masterpiece of careful, accurate observation; any ideas of a thick lunar atmosphere were dispelled, and Beer and Mädler showed that the Moon was both airless and changeless. Even so, there were dissentients. A German astronomer who rejoiced in the name of Franz von Paula Gruithuisen published drawings which, he said, showed a lunar city with 'dark gigantic

TYCHO AND KEPLER

The revolution in human outlook, started by Copernicus with his great book published in 1543, was continued — unintentionally — by the next major figure in the story, Tycho Brahe. I say 'unintentionally' because Tycho, a self-opinionated and haughty Danish nobleman, could never bring himself to believe that the Earth could move round the Sun. Instead, he perfected a sort of hybrid system, according to which the planets were in orbit round the Sun while the Sun itself moved round the Earth.

Yet Tycho was a superbly accurate observer. In 1572 he recorded a brilliant new star in the constellation of Cassiopeia, now known to have been a supernova, and decided to devote the rest of his life to astronomy. In 1576 he set up an observatory at Hven, an island in the Baltic, and remained there for twenty years, during which time he drew up much the best star catalogue ever compiled up to that time. All his work had to be done with the naked eye; telescopes were not invented until the first decade of the seventeenth century.

When Tycho left Hven, following a quarrel with the Danish court, he went to Bohemia to become Imperial Mathematician to Rudolph II, the Holy Roman Emperor. There he was joined by a young German mathematician, Johannes Kepler, who acted as his assistant. When Tycho died suddenly, in 1601, all his observations came into Kepler's possession, and Kepler used them well. As well as the star catalogue, there were extensive observations of the movements of the planets, particularly Mars. Kepler made up his mind to see whether he could, at last, find out the truth about the status of the Earth.

Up to that time it had always been assumed that the planets must move in circular paths, because the circle is the 'perfect' form, and presumably nothing short of absolute perfection can be allowed in the heavens. Yet so far as Mars was concerned, Kepler had to realise that Tycho's measurements could not be made to fit in with the concept of circular orbits. Finally Kepler solved the problem. The planets — including the Earth — do indeed move round the Sun, but their orbits are elliptical rather than circular. Once the great breakthrough had been made, Kepler was able to draw up his three famous Laws of Planetary Motion. They are as follows:

1 The planets move in elliptical orbits, the Sun being situated at one focus of the ellipse, while the other focus is empty.

2 The radius vector, or imaginary line joining the centre of the planet to the centre of the Sun, sweeps out equal areas in equal times. (In other words, a planet moves fastest when it is closest to the Sun, and slowest when it is at its greatest distance.)

3 The squares of the sidereal periods of the planets are proportional to the cubes of their mean distances from the Sun. (The sidereal period is the time taken to complete one orbit — 365¼ days in the case of the Earth.) The Third Law thus makes it possible to draw up a complete scale model of the Solar System, because the sidereal periods of the planets can be found by observation, and were already well known in Kepler's time; all that was needed to give the actual distances was one absolute measurement.

Kepler's Laws also apply to bodies such as comets, and to satellites moving round their primary planets. It is fair to say that these Laws, published between 1609 and 1618, really marked the overthrow of the Ptolemaic system, though the Church continued to oppose the new ideas, and not until the work of Newton in 1687 were all doubts finally removed.

It is true that the orbits of the principal planets are almost circular — the Earth's distance from the Sun does not vary by more than about 3,000,000 miles — but the difference is vitally important. Remember, however, that Kepler could never have made the discovery without having complete faith in Tycho's observations; and it is ironical that Tycho, who would have made such brilliant use of the telescope, missed its invention by less than ten years.

Driving across the Moon: Apollo 17. The astronauts were Eugene Cernan and Harrison Schmitt

ramparts', though in fact the area contains nothing more significant than low, haphazard ridges. Well before the middle of the nineteenth century the whole idea of advanced life on the Moon had been given up.

Vegetation was a different matter, and as recently as 1924 a famous American astronomer, W.H. Pickering, was still maintaining that certain dark patches in some of the craters indicated either low-type plants, or even swarms of insects or small animals. It was only with the coming of the space-ships, in our own time, that the Moon finally proved to be completely sterile.

Nobody has ever bettered the description of the lunar surface given by the second man on the Moon, Edwin Aldrin, on stepping out from the lunar module *Eagle* in July 1969: 'Magnificent desolation'. And I cannot do better than quote the words of the last man on the Moon, Eugene Cernan, when I talked to him ten years after his journey there in Apollo 17, in December 1972:

> Overall, everything was rather overshadowed by the view of the Earth itself. It's been said that the Moon is colourless, bland and unbeautiful, but that's not true; the Moon is beautiful and majestic, with its valleys and its towering mountains. Yet to me, the greatest realization was that of standing in the middle of a dim charcoal picture, looking back through the blackness of space — blackness even though it's filled with sunlight — and seeing the Earth, with all its life and colour. That meant more to me than stepping out on to the surface of the Moon.

Tycho Brahe

Tycho Brahe was a curious character. His life began on an unusual note, as he was kidnapped by his uncle; when he went to university he had part of his nose sliced off in a duel, and made himself a replacement out of gold, silver and wax (what happened when he caught a head-cold is not clear). When he established his observatory on Hven, in the Baltic, he was also made landlord of the island, and to say that the local inhabitants disliked him is to put it mildly. The observatory even had an adjacent prison, in which Tycho used to incarcerate those tenants who refused to pay their rents. The 'staff' of the observatory also included a pet dwarf, Jep, who was said to have the gift of second sight. Hven became very much of a scientific Mecca; one visitor was King James I.

THE LUNAR WORLD

Because the diameter of the Moon is more than one-quarter that of the Earth, it may be wise to regard the Earth-Moon system as a double planet rather than as a planet and a satellite. The origin of the Moon is still uncertain. The old theory that it broke away from the Earth, leaving the immense hollow now filled by the Pacific Ocean, has long since been discarded, and it is more likely that the two worlds were formed at the same time, in the same way, in the same region of space. The Moon's age seems to be very similar to that of the Earth (around 4.6 thousand million years). There is always the possibility that the Moon used to be totally independent, and was captured by the Earth early in the story of the Solar System; nobody really knows.

The Moon's condition differs from that of the Earth mainly because it is smaller, and has a much weaker pull of gravity; the escape velocity is a mere 1½ miles per second, as against 7 miles per second for the Earth, and this means that the Moon has been unable to hold on to any atmosphere. It retains a good deal of internal heat, however — it is not a dead, cold body, as was once believed — and there has been a tremendous amount of past activity there.

The principal surface features are the seas or 'maria', mountains and craters. The seas were never water-filled (in fact we are now sure that there never has been any water on the Moon) but were once filled with lava, so that their names are not entirely inappropriate. Most of the seas on the Earth-turned hemisphere of the Moon are connected, one notable exception being the Mare Crisium or Sea of Crises, which is not far from the north-east edge of the disk and is clearly identifiable with the naked eye.

The walled formations always known as craters are everywhere; they cluster thickly in the bright highlands, and are also found on the floors of the maria. They are basically circular, but break into each other and deform each other to such an extent that many of them are barely traceable. In size they range from vast enclosures over 150 miles in diameter down to tiny pits so small that from Earth they cannot be seen at all. Many of them have terraced walls and high central mountains or mountain groups. In general they have been named after famous scientists; Ptolemy, Galileo, Newton, Copernicus and Halley all have their own particular craters.

A crater is not a deep, steep-sided hole, such as a mine-shaft. Generally the wall rises to only a modest height above the outer landscape, while the floor is sunken. On average the slopes are fairly gentle, and there is no case of a central mountain reaching the height of the outer rampart. Some craters, notably Tycho in the southern uplands and Copernicus in the Mare Nubium (Sea of Clouds), are the centres of systems of bright streaks or rays, well seen only under a high angle of illumination — that is to say, when the Moon is not far from full.

The main lunar mountain ranges form the boundaries of the regular maria, so that they are not of the same type as our Himalayas; thus the Lunar Apennines and Lunar Alps form part of the border of the well-marked Mare Imbrium (Sea of Showers). Isolated peaks and clumps of hills are also common. Among the minor features are the domes, which are low swellings often capped by craterlets, and the rills (also known as rilles or clefts) which are depressions looking almost like the cracks in dried mud — though there has never been any mud on the Moon!

There has been endless argument about the origin of the Moon's craters. The majority view, particularly popular in the United States, is that they were produced by meteorite impacts in a concentrated cosmic bombardment between about 4,100 million and 3,800 million years ago, after which there was tremendous outpouring of volcanic material from below the crust, flooding the mare basins. The alternative theory,which I support, holds that the formations are of internal origin ('volcanic', if you like) so that they are basically of the same nature as our volcanic calderas. Certainly the distribution of the craters is not random; for example, when one crater breaks into another it is almost always the smaller formation which intrudes into the larger, which is an argument against chance bombardment. In any case, we may be sure that there are some impact craters on the Moon, and equally certainly there are volcanic structures; for instance, the chains of small craterlets seen in many areas can only be of internal origin.

The main active period in lunar history ended well over 2,000 million years ago, so that even a 'young' crater is ancient by terrestrial standards. The modern Moon is almost changeless. All that can be seen are occasional emissions of gas from below the surface, and even these so-called 'Transient Lunar Phenomena' (TLP for short) are very elusive.

The Apollo missions have confirmed that there is a loose upper layer or 'regolith', less than 50ft deep, below which is a layer of shattered bedrock overlying a series of layers of more solid rock. At a much deeper layer the rocks are hot enough to be melted, and there seems to be an iron-rich core with a diameter which may be as much as 1,000 miles. Despite the presence of this core, the Moon has no detectable overall

continued overleaf:

magnetic field, though there are certain regions of 'localised magnetism'. It has been suggested that there may once have been a general magnetic field which has now disappeared.

The boundary between the sunlit and 'night' hemispheres of the Moon is known as the terminator. When a large crater lies on or very close to the terminator it is spectacular, with shadow covering part or all of the floor; as the Sun rises over it, the crater becomes less prominent, and near full moon, when there are virtually no shadows at all, even a major crater may become hard to identify. Instead, the surface is dominated by the bright rays — chiefly from the craters Tycho and Copernicus — which are surface deposits, and are almost impossible to see under low lighting.

Though most craters are basically circular, the fact that the Moon always keeps the same face turned toward the Earth means that formations close to the edge or limb are very foreshortened. A circular crater appears elliptical, and in extreme cases it is very difficult to distinguish between a crater wall and a ridge. Until 1959, when the Russians sent their unmanned spacecraft Luna 3 on a 'round trip' and obtained photographs, we knew nothing definite about the far side of the Moon, though we guessed — correctly — that it must be just as cratered and just as barren as the area we have always known.

THE NAMES OF THE MAIN LUNAR 'SEAS' ARE AS FOLLOWS:

Sinus Æstuum:	The Bay of Heats
Mare Australe:	The Southern Sea
Mare Crisium:	The Sea of Crises
Mare Foecunditatis:	The Sea of Fertility
Mare Frigoris:	The Sea of Cold
Mare Humboldtianum:	Humboldt's Sea
Mare Humorum:	The Sea of Humours
Mare Imbrium:	The Sea of Showers
Sinus Medii:	The Central Bay
Mare Nectaris:	The Sea of Nectar
Mare Nubium:	The Sea of Clouds
Mare Orientale:	The Eastern Sea
Oceanus Procellarum:	The Ocean of Storms
Sinus Roris:	The Bay of Dews
Mare Serenitatis:	The Sea of Serenity
Sinus Iridum:	The Bay of Rainbows
Lacus Somniorum:	The Lake of the Dreamers
Palus Somnii:	The Marsh of Sleep
Mare Tranquillitatis:	The Sea of Tranquillity
Mare Vaporum:	The Sea of Vapours

AMONG SOME OF THE LARGE WALLED FORMATIONS ARE:

Name	*Diameter (miles)*	*Comments*
Albategnius	80	Companion to Hipparchus.
Alphonsus	80	Member of the Ptolemæus chain. Low central peak.
Anaxagoras	32	Northern area; ray centre.
Archimedes	50	On Mare Imbrium. Regular, dark-floored.
Aristarchus	23	On Oceanus Procellarum. Brightest feature on the Moon.
Aristillus	35	Member of the Archimedes group.
Aristoteles	60	Pair with Eudoxus; edge of Mare Frigoris.
Atlas	55	N of Lacus Somniorum, pair with Hercules.
Autolycus	22	Member of the Archimedes group.
Bailly	182	'Field of ruins; near the SW limb.
Bullialdus	31	Fine crater, on the S part of the Mare Nubium. Central peak and terraced walls.
Casatus	65	Southern uplands; intrudes into Klaproth.
Catharina	55	Member of the Theophilus chain.
Clavius	144	Huge enclosure in S uplands; continuous walls, with a string of craters across its floor.
Cleomedes	78	Mare Crisium area.
Copernicus	60	Great ray-crater. Massive walls; central peaks.
Cyrillus	60	Member of the Theophilus chain.
Doppelmayer	42	Bay in Mare Humorum; 'seaward' wall destroyed.
Einstein	99	Large formation very near the SW limb.
Endymion	73	Dark-floored formation near Mare Humboldtianum.
Eratosthenes	38	Very deep and well-formed; lies at the end of the Apennines chain of mountains.
Fracastorius	60	Great bay in the Mare Nectaris.
Fra Mauro	50	Ruined crater on Mare Nubium; forms a trio with Bonpland and Parry.

Furnerius	80	Near the SE limb; member of the Petavius chain.
Gassendi	55	Fine crater; N edge of Mare Humorum, with a low central peak and rills on its floor.
Grimaldi	120	Very dark-floored enclosure near the W limb.
Hevel	76	Near Grimaldi; convex floor, low central peak.
Hipparchus	90	Semi-ruined; closely east of Ptolemæus.
Kepler	22	Major ray-centre on Oceanus Procellarum.
Langrenus	85	W edge of Mare Fœcunditatis. Massive walls, with a central peak.
Longomontanus	90	Clavius area; complex walls.
Macrobius	42	Mare Crisium area.
Maurolycus	68	Southern uplands, near Stöfler.
Menelaus	20	Brilliant crater in the Hæmus Mountains.
Petavius	105	Near the NE limb; massive walls, central peak, and a prominent floor-rill.
Piccolomini	50	End of the Altai Scarp.
Pitatus	50	Low-walled; on the S edge of Mare Nubium.
Plato	60	Dark-floored; lies at the edge of Mare Imbrium.
Plinius	30	Fine crater, between Mare Serenitatis and Mare Tranquillitatis.
Proclus	18	Brilliant crater at the edge of Palus Somnii.
Ptolemæus	92	Major formation near the centre of the Moon's disk; smooth floor, no central peak.
Purbach	71	Member of the Walter-Regiomontanus chain.
Pythagoras	70	Fine crater very near the NW limb.
Regiomontanus	80x65	Irregular; between Walter and Purbach.
Riccioli	99	Near Grimaldi; very dark area on the floor.
Scheiner	70	Near Clavius; pair with Blancanus.
Schickard	126	Great enclosure in the far SW region.
Taruntius	37	Concentric crater on Mare Fœcunditatis.
Thebit	37	Edge of Mare Nubium; broken by Thebit A, which is itself broken by Thebit F.
Theophilus	63	Splendid crater on the edge of Mare Nectaris; high, terraced walls and massive central peak.
Triesnecker	14	On Mare Vaporum; associated with a complex system of rills.
Tycho	52	Great ray-crater in the southern uplands. Terraced walls, central peak.
Vendelinus	103	Petavius chain; low, broken walls.
Walter	80	Outside Mare Nubium; massive walls, central peak.
Wargentin	55	Near Schickard. Lunar plateau,with a raised interior.

AMONG THE MAIN MOUNTAIN RANGES ARE:

Alps	N border of the Mare Imbrium.
Altai Scarp	SE of Mare Nectaris.
Apennines	Part of the border of the Mare Imbrium.
Carpathians	Bordering Mare Imbrium, to the south.
Caucasus	Separating Mare Serenitatis from Mare Imbrium.
Hæmus	Part of the S border of Mare Serenitaris.
Harbinger	Clumps of hills in the Aristarchus region.
Jura	Bordering Sinus Iridum.
Pyrenees	Clumps of hills bordering Mare Nectaris, to the E.
Riphæans	Short range in the Mare Nubium.
Spitzbergen	Mountain clump in Mare Imbrium, near Archimedes.
Straight Range	Very regular range near Plato, in Mare Imbrium.
Taurus	Mountain clumps E of Mare Serenitatis.

Neighbours

Opposite:
Mercury, from Mariner 10. Superficially the surface is very like that of the Moon, with mountains, valleys and craters, though there are significant differences in detail

Neighbours always interest us. Not all of them are friendly, but all have their own special characteristics. In the universe, our particular neighbours are the members of the Solar System, and from the very earliest days men have wondered whether there can be anyone 'up there' looking at our own Earth. In this respect the Moon has failed us; so what about the planets?

Planets, remember, are bodies of the same basic nature as the Earth, shining only by reflecting the rays of the Sun. They are divided into two groups. First come the small inner planets: Mercury, Venus, the Earth and Mars. Next there is the wide gap filled by the thousands of minor planets or asteroids, and beyond we meet the giants Jupiter, Saturn, Uranus and Neptune, plus the small and perplexing Pluto. They make up a varied family.

If we are to find life, what sort of a world must we seek? Perhaps we must first define what is meant by 'life'. Beings such as ourselves require a reasonably even temperature, the right kind of atmosphere,

Rotation of Venus. In these Mariner pictures the feature 'arrowed' is clearly identifiable. The upper clouds have a rotation period of 4 days, as against 243 days for the globe!

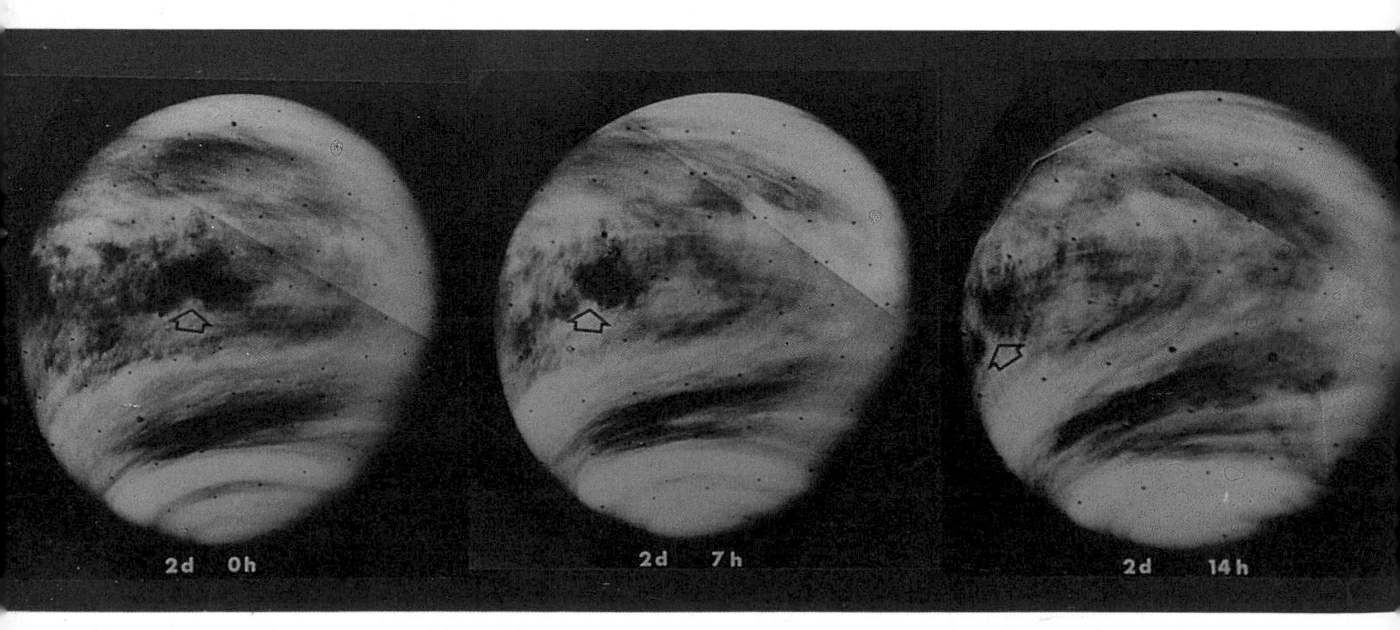

and a plentiful supply of water. If any of these essentials are missing, then we must look elsewhere, and most of the bodies of the Solar System are ruled out without further ado.

The Moon is a case in point. The low escape velocity of 1½ miles per second means that any past lunar atmosphere has long since leaked away into space, and the same is true of the first of the inner planets, Mercury, which has the added disadvantage of being very close to the Sun; at noon during its long 'day' the Mercurian rocks are hot enough to melt a tin kettle, while during the equally long night the temperature falls far below anything we experience on Earth. None of the asteroids can retain any trace of air, and neither can the satellites of the giant planets with the sole exception of Titan, in the system of Saturn. (I exclude 'atmospheres' which are so thin that we would normally class them as being no better than a vacuum.) The

THE PLANETS

All the planets have their own particular features of special interest. Data are as follows:

Planet	Mean distance from Sun (miles)	Sidereal Period ('year')	Rotation Period (equatorial)	Diameter in miles (equatorial)
Mercury	36,000,000	88 days	58.65 days	3,033
Venus	67,000,000	224.7 days	243.2 days	7,523
Earth	92,957,000	365.3 days	23hr 56min 4sec	7,926
Mars	141,500,000	687 days	24hr 37min 23sec	4,218
Jupiter	483,300,000	11.9 years	9hr 50min 30sec	89,400
Saturn	886,100,000	29.5 years	10hr 39min	74,900
Uranus	1,783,000,000	84.0 years	17hr 14min	31,770
Neptune	2,793,000,000	164.8 years	16hr 3min	31,410
Pluto	3,667,000,000	247.7 years	6 days 9hr17min	1,444

Of these, the first four have solid surfaces and are often called the 'terrestrial' planets; Jupiter, Saturn, Uranus and Neptune have surfaces made up of gas, with relatively small solid cores. Pluto, which has a much more eccentric orbit than those of the other planets, seems to be in a class of its own; its 'satellite', Charon, has over half the diameter of Pluto itself.

The bright planets have been known since ancient times. Of the rest, Uranus was discovered in 1781, Neptune in 1846 and Pluto in 1930. Uranus can just be seen without optical aid; binoculars will show Neptune, but Pluto is invisible unless a telescope is used.

Slight irregularities in the motions of Mercury led nineteenth-century astronomers to believe that there might be another planet moving in an orbit even closer to the Sun. It was even given a name: Vulcan. Obviously it would be very hard to see, and the only chances of finding it seemed to be either during a total solar eclipse or else at a time when it passed in transit across the face of the Sun. All attempts to identify it failed, and the movements of Mercury have now been satisfactorily explained by Einstein's theory of relativity. On the other hand there is a strong possibility that a new outer planet exists, moving far beyond the orbits of Neptune and Pluto.

It used to be thought that the planets were pulled off the Sun by the action of a passing star, between 4½ and 5 thousand million years ago. If this had been true, then Solar Systems would have been uncommon in the Galaxy, simply because the stars are widely spaced; 'close encounters' must be very rare indeed. Mathematical objections have led astronomers to abandon this theory, and it now seems certain that the planets were formed by condensing out of a cloud of dust and gas associated with the youthful Sun.

MERCURY

Though Mercury has been known since very early times, there must be many people who have never seen it. It always remains close to the Sun in the sky, and is visible with the naked eye only when very low in the west after sunset or very low in the east before sunrise. It is a quick mover, and was fittingly named after the fleet-footed messenger of the gods.

Very little can be seen on its surface with Earth-based telescopes. All that can be made out are a few darkish patches, and it is now known that all maps of it drawn before the Space Age were very inaccurate. It was also believed that the rotation was captured or synchronous — that is to say, equal to the revolution period of 88 days, so that Mercury would keep the same face turned permanently toward the Sun. There would be a region of everlasting day and another region from which the Sun would never be seen, with only a narrow 'twilight zone' in between over which the Sun would bob up and down over the horizon. If this had been the case, then the dark side of Mercury would have been bitterly cold, perhaps even colder than remote Pluto.

Then, in 1962, American astronomers measured the long-wavelength radiations from Mercury, and found that the dark side was much warmer than it would be if it never received any sunlight. We now know that the true rotation period is 58.65 Earth days, or two-thirds of a Mercurian 'year'.

The low escape velocity of 2½ miles per second means that Mercury has virtually no atmosphere, so that in many ways it is not unlike the Moon. Searches for a satellite have been unsuccessful, and we are now sure that Mercury, like Venus, is a solitary traveller in space.

Most of our direct knowledge of the planet has been drawn from one probe, Mariner 10. It was launched in November 1973, and bypassed Venus in the following February before going on to rendezvous with Mercury in late March 1974. It obtained excellent close-range pictures, as well as a great deal of general information, and then continued in orbit round the Sun. It made a second encounter with Mercury in September 1974 and a third in March 1975, but by then its equipment was starting to fail, and contact with it was finally lost on 24 March. No doubt it is still moving round the Sun, and still making regular close approaches to Mercury, but we have no hope of finding it again.

Mariner 10 showed that Mercury is mountainous and cratered. There is one huge ringed formation, known as the Caloris Basin, but otherwise there are no 'seas' similar to the lunar maria; on the other hand there are craters of all kinds, some of them very large, together with mountains, cliffs and relatively smooth plains. The craters have been named after persons who have made major contributions to human culture, such as Chopin, Beethoven, Dickens, Rubens and Mark Twain. Some of the craters have terraced walls and central peaks, and ray-centres are also to be found.

Mercury has a weak but appreciable magnetic field. Overall, the globe is as dense as that of the Earth, and it seems that there is a relatively large iron-rich core, probably larger than the entire body of the Moon.

Unfortunately the same regions of Mercury were in sunlight during each Mariner pass, so that we have proper maps of less than half the total surface. There seems no reason to believe that the remaining regions are basically different, but to find out for certain we will have to wait for a new space-probe, and there seems no chance that this will be launched yet awhile. It is hardly necessary to add that any form of life there is completely out of the question.

four giant planets have gaseous surfaces, and are mainly liquid inside, with only relatively small solid cores. Obviously, then, our choice is limited from the outset.

There is one loophole in this argument. Suppose that we can turn to an entirely alien form of life, which can manage very well without air or water, and in temperatures which to us would be fatal? Science fiction writers are very fond of such aliens, and call them BEMs or Bug-Eyed Monsters, but there is no evidence that they can exist, and a great deal of evidence that they do not. So for the moment it is sensible to restrict ourselves to 'life as we know it', and this brings us on to our two main candidates, Venus and Mars.

Of the two , Venus is the closer. It can come within 25,000,000 miles of us, which is only about a hundred times as far away as the Moon, and in size and mass it and the Earth are near-twins. Represent the Earth by a snooker ball, and Venus will be another ball so like it that Steve Davis could quite happily use both for play. Therefore, it would be logical to assume that Venus and the Earth ought to be in much the same condition, allowing for the fact that Venus is just over 20,000,000 miles closer to the Sun than we are. Yet nothing could be further from the truth.

There is plenty of atmosphere; in fact there is too much, because the clouds never clear, and nobody has yet had a direct view of the planet's surface. Telescopic observation could tell us very little, and before 1962 there were two main theories.—either Venus was a wild, fiercely hot dust-desert, or else it was a comparatively pleasant place, covered largely with swamps or ocean. Even the length of Venus' 'day' was unknown, though the 'year' (the time taken to go once round the Sun) had been fixed very accurately at 224.7 Earth-days.

One intriguing suggestion was made by Franz von Paula Gruithuisen (yes, the same enthusiast who reported 'dark gigantic ramparts' on the Moon). As Venus moves along in its path, it shows

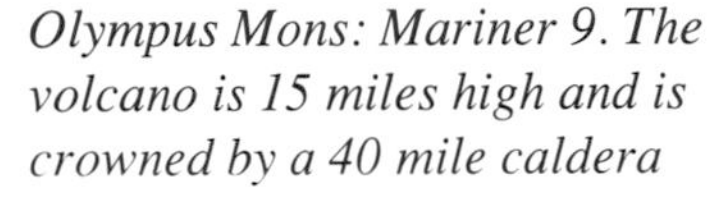

Olympus Mons: Mariner 9. The volcano is 15 miles high and is crowned by a 40 mile caldera

The Painted Desert of Arizona. Lowell compared this with a Martian landscape. Admittedly it is easy to agree that the two are somewhat alike

Mariner 2, the first successful interplanetary probe, which surveyed Venus in 1962

regular phases or changes of shape, similar to those of the Moon and for much the same reason. During the crescent stage, the non-sunlit side can sometimes be seen shining dimly. This appearance is very common with the crescent moon (it is often called 'the Old Moon in the Young Moon's arms'), but there is no mystery here; it is due to light reflected by the Earth on to the lunar surface. But this cannot explain the so-called Ashen Light of Venus, if only because Venus has no moon; like Mercury, it is a solitary traveller in space.

The Ashen Light is not easy to see. You need an adequate telescope, special equipment, and a good deal of patience and luck. However, it has been reported so many times, and by so many experienced observers, that its reality is not in doubt. Gruithuisen, of course, had no doubts at all, and he also had a solution. He pointed out

VENUS

Venus, the next planet in order of distance from the Sun, is as different from Mercury as it could possibly be. It is more brilliant than any object in the sky apart from the Sun and the Moon, partly because it is relatively close to us — at its nearest, only about a hundred times as far away as the Moon — and partly because its cloudy atmosphere makes it very reflective. At its best, in the west after sunset or in the east before dawn, it can cast strong shadows.

Beautiful though it looks as seen with the naked eye, we have to admit that Venus is a telescopic disappointment. All that can be made out is the characteristic phase; where we might have expected to see mountains and valleys, all that we can glimpse are vague dusky shadings which are almost impossible to define, and which shift and change quickly. The clouds on Venus never clear; nobody has ever had a direct view of the surface.

Another problem concerned the length of the rotation period. Venus takes 224.7 days to complete one journey round the Sun, but what about its 'day'? The estimated period was of the order of a month, but this was little more than a guess.

The first really reliable information came from Mariner 2, in 1962, which bypassed the planet and told us that the surface is fiercely hot. The rotation period proved to be 243 Earth-days, longer than Venus' 'year', and to make the situation even stranger Venus rotates from east to west, so that if it were possible to stand on the surface and look at the Sun it would be found that sunrise occurred in the west and sunset in the east. Venus is, in fact, an 'upside-down' world, for reasons which remain unknown.

The next missions were dispatched by the Russians, who did their best to parachute their probes down through Venus' atmosphere and make controlled landings. Not surprisingly, they had several failures, but in 1970 they achieved a notable triumph; Venera 7 touched gently down, and managed to transmit for over twenty minutes before being put out of action by the intensely hostile conditions. The picture showed a rock-strewn, dimly lit landscape which seemed just as unfriendly as the surface of Mercury.

Since then many more space-probes have been sent to Venus, some of which have landed while others have been put into orbits round the planet to map the surface by radar. The atmosphere is made up chiefly of carbon dioxide, and the ground pressure is about ninety times that of the Earth's air at sea-level; the temperature is so high that the rocks glow dull orange, and the clouds contain large quantities of corrosive acid. Well over half the surface is covered by a vast, rolling plain, with craters here and there, while there are two major highland areas, picturesquely named Ishtar and Aphrodite. Mountains, valleys and plateaux have been identified, and there seems every chance that there are active volcanoes, with almost continuous thunder and lightning.

Why is Venus so different from the Earth? The answer must lie in its lesser distance from the Sun. It is thought that in the early story of the Solar System, the Sun was less luminous than it is now, so that Venus and the Earth may have started to evolve along similar lines. When the Sun became more powerful, the oceans of Venus evaporated, and there was what may be called a 'runaway greenhouse' effect, so that in a relatively short time Venus was turned into the furnace-like world of today. Any life which may have gained a foothold there would have been ruthlessly snuffed out. If this is true, then we must regard Venus as a rather sad world. Certainly there is no chance of our being able to visit it in the foreseeable future.

that the Light had been observed in 1759 and again in 1806, an interval of 47 terrestrial or 76 Venus years, and went on to say: 'We assume that some (Venus) Alexander or Napoleon then attained universal power. If we estimate that the ordinary life of an inhabitant of Venus lasts 130 Venus years, which amounts to 80 Earth years, the reign of an Emperor of Venus might well last for 76 Venus years. The observed appearance is evidently the result of general festival illumination in honour of the ascension of a new emperor to the throne of the planet.'

Later on, Gruithuisen modified this idea somewhat. Instead of a coronation, he suggested that the Light might be due solely to the burning of large stretches of jungle to produce new farm land, and added that 'large migrations of people would be prevented, so that possible wars would be avoided by abolishing the reason for them. Thus the race would be kept united.'

Not surprisingly, Gruithuisen's theories were met with a certain amount of scepticism. Later explanations of the Light ranged from phosphorescent oceans to a self-luminous atmosphere, but nowadays it seems much more likely that the glow is due to electrical storms in the planet's upper atmosphere, not unlike our auroræ or polar lights.

Despite Gruithuisen, the prospect of a reasonably welcoming world persisted. Svante Arrhenius gave a very vivid picture of the planet's surface, which he believed to be in the same state as that of the Earth over 200 million years ago — in the so-called Carboniferous Period, when the coal forests were being laid down and even the ferocious dinosaurs lay in the future. He concluded that:

> 'everything on Venus is dripping wet . . . A very great part of the surface is no doubt covered with swamps, corresponding to those of the Earth in which the coal deposits were formed, except that they are about 30°C warmer . . . Only low forms of life are represented, mostly no doubt belonging to the vegetable kingdom. The vegetative processes are greatly accelerated by the high temperature. Therefore, the lifetime of organisms is probably short . . . The temperature at the poles of Venus is probably somewhat lower than the average temperature on the planet. The organisms there should have developed into higher forms than elsewhere, and progress and culture, if we may so express it, will gradually spread from the pole toward the equator. Later, the temperature will sink, the dense clouds and the gloom disperse, and some time, perhaps not before life on Earth has reverted to its simpler forms or has even become extinct, a flora and a fauna will appear, similar in kind to those which now delight our human eye, and Venus will then indeed be the 'Heavenly Queen' of Babylonian fame, not because of her radiant lustre alone, but as the dwelling-place of the highest beings in our Solar System.'

One immediate objection to this picture was the apparent lack of water vapour in Venus' upper atmosphere, but it was impossible to tell what might lie lower down. Then, in the 1930s, studies showed

Venus: the Ashen Light, 1988 (15in reflector). The Light has been very much enhanced – otherwise it would not have been visible in the drawing

Oceans of oil?

In 1955 a rather novel picture of Venus was proposed by Sir Fred Hoyle. He believed the planet to have oceans of oil, and concluded that 'Venus is probably endowed beyond the dreams of the richest Texas oil-king'.

Magellan to Venus

On 4 May 1989 NASA launched Magellan, a probe designed to make a radar map of almost the whole of Venus. Magellan operated until well into 1993, and was a great success, showing lava flows, craters, huge volcanoes and structures of all kinds. The scientific instrument used for the mapping was SAR (Synthetic Aperture Radar) which can penetrate the thick cloud cover.

that the atmosphere contains a great deal of the heavy gas carbon dioxide, and this led to another picture, due to two American astronomers, Donald Menzel and Fred Whipple. Their Venus was marine, with only isolated islands here and there. Since the carbon dioxide in the atmosphere would have fouled the water, the result would have been seas of soda-water, though I remember pointing out that the chances of finding any Scotch to mix with it did not seem bright!

If Venus had really been like this, then there was an interesting follow-up. So far as we can tell, life on Earth appeared in our warm oceans several thousands of millions of years ago, and slowly evolved — single-celled organisms into multi-celled creatures, then fishes, amphibians, reptiles, mammals, and finally you and me. A marine Venus might well be a replica of the Earth in past ages, and in this case it was fair to assume, as Arrhenius had done, that things might follow the same course. It was even suggested that Venus might be an Earth in the making, with Mars representing an Earth in decline.

Until we had some definite information to guide us, there was really no reason to dismiss Venus as lifeless, though the high temperature and the dense, carbon-dioxide-rich atmosphere ruled out anything in the way of civilisation. Alas for these intriguing ideas! In 1962, Mariner 2, the first successful interplanetary spacecraft, flew past Venus and sent back information which was decidedly

MARS

Mars, the Red Planet, has always been of special interest to us. It is smaller than the Earth, and around 50,000,000 miles further away from the Sun. It never approaches much closer than 35,000,000 miles of the Earth and therefore even our best telescopes can give us no better a view than we can obtain of the Moon with low-power binoculars.

Moreover, it comes to opposition only in alternate years; thus there were oppositions in 1986, 1988 and 1990, but not in 1987 or 1989. At its best Mars is very brilliant, and can even outshine Jupiter, though when furthest from the Earth it looks exactly like a fairly bright reddish star.

Mars takes 687 Earth-days to complete one orbit. This is equivalent to 669 Martian days or 'sols', since the planet spins round rather more slowly than we do. The orbit is much less circular than that of the Earth, and this means that the amount of energy it receives from the Sun varies considerably over a Martian 'year', but the tilt of the axis is almost the same as that of the Earth, and the seasons are of the same basic type, though they are of course much longer.

The Martian 'air' is very thin, because the planet's pull of gravity is too weak to retain an atmosphere of the same kind as that of the Earth. The main constituent is carbon dioxide, and the ground pressure is below 10 millibars everywhere, so that the atmosphere is much more rarefied than that of the Earth at three times the height of Everest. The winds can be quite rapid, but have little force; on the other hand there are frequent dust-storms, which may cover large areas of the planet and hide the surface completely.

Much of the surface is 'desert', but the dark markings, now known to be simply areas from which the reddish covering has been scoured away, are permanent; the most prominent of them, such as the V-shaped Syrtis Major, can be seen with a small telescope when Mars is well placed. They are not old sea-beds, as used to be believed, and indeed some of them — including the Syrtis Major — are lofty plateaux.

The polar caps wax and wane with the seasons, so that they are very extensive in Martian winter and almost vanish near midsummer. They are genuinely

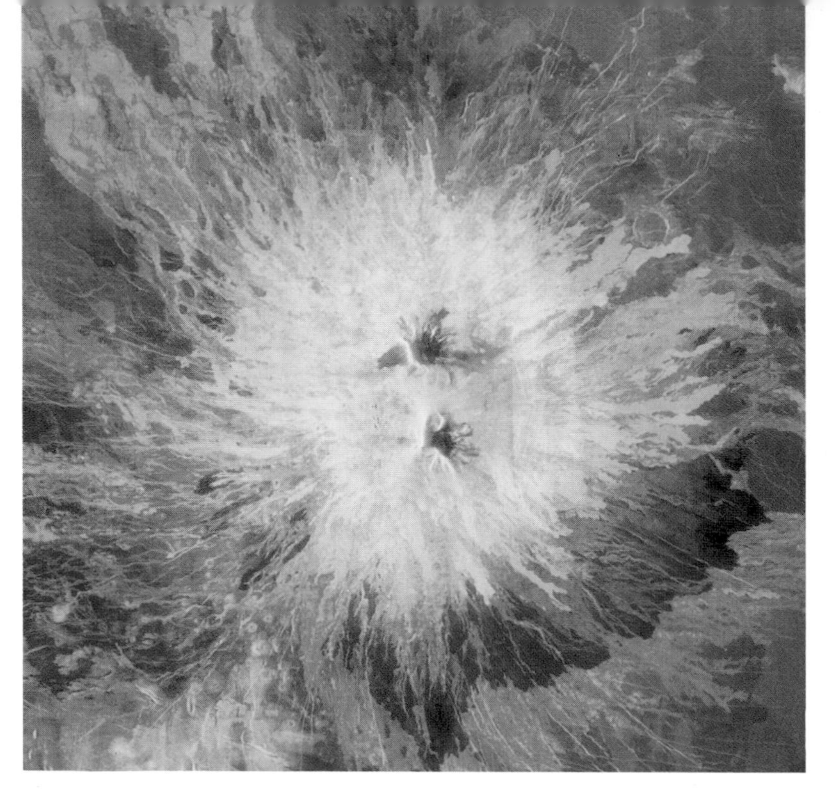

False colour image of volcano Sapas Mons; Venus – Magellan, September 1991. Sapas Mons is 250 miles across and almost a mile high. The summit has two smooth mesas, and there are lava flows along the flanks.

Martian craters seen from Earth

It is usually said that the craters on Mars cannot be seen from Earth, but this may not be true, because they were reported twice in pre-Space Age times: by Edward Emerson Barnard, using the 36in Lick refractor in 1892, and by John Mellish, with the 40in Yerkes refractor in 1915. Yet neither set of observations was ever published. Barnard witheld his because he thought that 'people would laugh at him', while Mellish's drawings were destroyed during a disastrous fire at his observatory.

unwelcome. The planet was much too hot for terrestrial-type life, and there could be no seas. Later probes made things even worse. The surface temperature is not far short of 1000°F, the atmospheric pressure is crushing, and the clouds contain large amounts of

icy, and are many feet thick, but the water ice is mixed with 'dry ice'— solid carbon dioxide. The overall temperature of the planet is low. Even at noon on the equator, in summer, a thermometer would barely rise to 50°F, and the nights are bitterly cold, so that all in all the climate of Mars is far from welcoming.

The first successful Mars probe, Mariner 4, bypassed the planet in July 1965, and sent back some rather unexpected information. Instead of being smooth or at best gently undulating, the surface proved to be cratered, so that Mars was more like the Moon than like the Earth. Later space-craft showed that as well as craters there are deep valleys, old riverbeds, and towering volcanoes, one of which — Olympus Mons, or Mount Olympus — rises to at least 25,000ft above the outer landscape. Evidently there has been a great deal of activity on Mars in the past, and we cannot be sure that the volcanoes are dead even yet.

In 1975 two American spacecraft, the Vikings, made controlled landings on Mars. Viking 1 came down in the ochre plain of Chryse, while Viking 2 followed some weeks later in the more northerly region known, rather inappropriately, as Utopia. Both were able to analyse the atmosphere, measure the winds and temperatures, and carry out a determined search for any signs of life. Rather to the general regret, no life was found, and it is now probable that Mars is sterile — at least at the present epoch. Whether it has always been so is quite another matter, but even if life appeared there it is not likely to have had enough time to evolve very far before the conditions became too hostile.

There are two satellites, Phobos and Deimos, both discovered by Asaph Hall in 1877. Both are small, and irregular in shape. Phobos has a longest diameter of less than 20 miles, Deimos less than 10, so that neither would be of much use as a source of illumination during the Martian night; Deimos would look like nothing more than a rather large, dim star. Phobos is very close to Mars — its distance from the surface is only about the same as that between London and Aden — and it has a revolution period of only 7½ hours, so that it completes three orbits in every Martian 'day'. To an observer on the planet, therefore, Phobos would appear to rise in the west, gallop across the sky, and set in the east only 4½ hours later. Very probably Phobos and Deimos are not true satellites, but merely ex-asteroids which were captured by Mars in the remote past.

Martian years

When we set up our first colonies on Mars, which will probably be at some time during the twenty-first century, we will have to work out a Martian calendar, bearing in mind that there will be 669 'days' or sols in every 'year'. My own idea is to divide the Martian year into 18 months, each of 37 sols. This makes 666 sols. We need 669, so an extra sol can be tacked on to Months 6, 12 and 18, giving them 38 sols instead of 37. This should work reasonably well, but no doubt the United Nations will produce something much more complicated.

Incidentally, the north pole star of Mars is not Polaris, but a much brighter star, Deneb in Cygnus (the Swan).

Another view

In 1944 a young science writer, Donald Lee Cyr, published a book in which he forecast craters on Mars, though he believed them to be impact structures and that the canals were fertility tracks left by animals migrating from one crater to another.

Martian bees?

In 1950 a well-known science writer named Gerald Heard announced that there really were Martians, far more intelligent than men — but they were only 2in long, and looked like bees.

sulphuric acid. Whether we will ever visit Venus I do not know, but it will not be in my time or yours. With regret, the 'Venusians' have been returned to the story-books.

Beyond the orbit of the Earth, we come to the red planet Mars. Though it is named after the God of War, it is much the least unfriendly of our neighbours, and up to less than thirty years ago it was generally believed that there was life there. Going back a few decades further, there were eminent astronomers who believed that they had seen signs of intelligent activity in the form of a vast, planet-wide system of canals. Only in our own time have the 'Martians' too been relegated to the world of myth.

I had my first telescopic view of Mars at the age of seven. I could see the reddish disk, some dark markings, and a white cap covering what was in fact the planet's south pole. Since then I have had the advantage of using some of the world's best telescopes, and Mars never fails to fascinate me. The reddish-ochre tracts are indeed deserts, though they are not like our Sahara; instead of being sandy wastes, with oases, palm-trees and strolling camels, they are covered with reddish 'dust', and they are bitterly cold instead of fiercely hot. Where we went wrong, until quite recently, was in believing that the dark patches were either seas or else old sea-beds filled with vegetation. It took the first Mars probe, Mariner 4 of 1965, to show us the error of our ways.

Because Mars is intermediate in size between the Earth and the Moon, with an escape velocity of just over 3 miles per second, we would expect a thin but appreciable atmosphere, and this is precisely what we find. From an early stage, the belief in Martians was widespread, and it was suggested that there might be possibilities of getting in touch with them. Karl Friedrich Gauss, one of the great nineteenth-century mathematicians, wanted to produce geometrical patterns in Siberia by planting pine-trees in suitable positions, reasoning that the sudden appearance of, say, an isosceles triangle would tell the Martians that we were trying to attract their attention. An Austrian astronomer, Carl von Littrow, went one better, transferring his scheme to the Sahara Desert and replacing Gauss' pine-trees with fire-filled ditches. But the pinnacle of invention was reached in 1869 by the French inventor Charles Cros, in his book *Moyens des Communications avec les Planètes*. His idea was to build a very long-focus mirror and concentrate the Sun's rays upon the Martian deserts, scorching the surface just as a burning-glass will scorch a piece of paper. He went on to add that by judicious swinging of the mirror it would even be possible to write words.

I have often wondered what words he intended to transmit, and it is difficult to judge how the Martians would have reacted to strange messages suddenly appearing in their deserts, but the experiment was never actually tried. Cros bombarded the French Government with letters, memoranda, petitions and pamphlets, but officialdom refused to co-operate, and eventually Cros gave up in disgust. Three years after Cros' abortive campaign came the first systematic observations of the Martian canals, and the possibility of intelligent life there began to be taken very seriously indeed.

Giovanni Virginio Schiaparelli was an Italian astronomer with an excellent reputation. In 1877 he began to map Mars from his observatory in Milan; he had an excellent telescope, and it is fair to say that his charts were more detailed than any previously drawn. Across the ochre deserts he drew numbers of straight, artificial-looking streaks, which he called *canali* (this is Italian for 'channels', but inevitably the translation into English converted them into canals, and canals they remained).

When Schiaparelli first published his drawings he met with a good deal of scepticism, because for some years nobody else managed to see the canals at all. The situation was made even more curious by Schiaparelli's account of the phenomenon of 'gemination', or twinning; a single canal might suddenly be replaced by a pair of streaks, strictly parallel. It was tempting to suggest that the canals were artificial, and had been built to convey water from the polar ice-caps through to the planet's equator; by that time it had become fairly clear that the Martian atmosphere was too thin for any major seas to exist on the surface.

In 1886 came what seemed, at the time, to be full confirmation of the canal network. Using a very powerful telescope at the Observatory of Nice, two French observers, Perrotin and Thollon, published a chart which was as singular as anything of Schiaparelli's, and before long the canals became thoroughly fashionable. They were described as black, well-defined, and easy to see; where one canal crossed another there was a sort of blob or 'oasis', presumably a centre of population. Canals were seen not only across the deserts, but across the dark regions as well.

Enter Percival Lowell, supreme champion of the canal theory, who built a major observatory at Flagstaff in Arizona mainly to observe the planet. Lowell equipped his observatory with a 24in refracting telescope which is still one of the finest in the world, and used it consistently from 1896 until his death in 1916. I can personally vouch for the quality of the telescope, because during my Moon-mapping period I made a large number of observations with it, and 'took time off' to look at Mars as well.

Lowell was a remarkable man. He was a fine speaker, a fine writer, a great benefactor of astronomy, and a skilled mathematician. The only thing he was *not*, unfortunately, was a good observer, and some of his drawings of Mars showed an amazingly complex network which was rather unkindly likened to a spider's web. He was utterly convinced that Mars was the site of an advanced civilization which was doing its best to survive upon a planet which was desperately short of water, and in 1906 he wrote as follows:

> That Mars is inhabited by beings of some sort or other we may consider as certain as it is uncertain what those beings may be. . . .The first thing that is forced on us in conclusion is the necessarily intelligent and non-bellicose character of the community which could thus act as a unit throughout its globe.

Eccentric calendars

Unusual calendars have been proposed for our own world from time to time. One of the oddest was the Calendar of the French Revolution, introduced in France in 1792. The hour was about twice as long as the normal hour, and there were twelve months of thirty days each; the beginning of the year was the autumn equinox. At the end of the year there were five (or, in Leap Years, six) holiday days, while the days of the month were divided into three groups of ten. This was all very well, but it meant that workers had only one day off in ten rather than one in seven, and there was also the slight disadvantage that nobody outside France used the new calendar. It was finally abandoned in 1806.

Much more recently there has been an even more remarkable scheme, introduced in 1989 by the left-wing Leeds City Council, who have decreed that their minions must convert to Metric Time. Gone are our normal sixty-minute hours; instead, the hour is divided into units of six minutes each, so that, for example, 10.16pm old time becomes 10.1 Leeds Time; a quarter past eleven becomes 11.25, and so on.

Mr Steven Kilburn, acting office manager of the Department of Technical Services of the Leeds City, was in no doubts. 'The new system is simplicity in itself,' he claimed. To emphasise this, let me quote from the official Council memorandum: 'An individual may wish to book the time using two decimal figures, and the system will cater for this. However, it is generally accepted that one decimal place is an adequate lowest common denominator.'

Somehow, I have a feeling that Leeds Time will not even last for as long as the Calendar of the French Revolution did!

Novel suggestions

In 1727 Jonathan Swift wrote *Voyage to Laputa,* one of the travels of the immortal Dr Lemuel Gulliver. Laputa was an airborne island (the original flying saucer!) and its keen-eyed astronomers had discovered two satellites of Mars, one of which revolved round the planet in only ten hours. This is less than the length of the Martian 'day', which was reasonably well-known even in Swift's time. In

continued opposite

> War is a survival among us from savage times and affects now chiefly the boyish and unthinking element of the nation. The wisest realise that there are better ways for practising heroism and other and more certain ends of insuring the survival of the fittest. It is something a people outgrow.

Lowell was careful to add that 'to talk of Martian beings is not to talk of Martian men', but it was fair to say that if his drawings of the planet had been accurate, then Mars would have been inhabited. Yet other astronomers either failed to see the canals at all, or else drew them as vague, ill-defined patches, not in the least artificial in aspect. This did not deter Lowell's disciples; one of them, W. B. Housden, even worked out the exact design and power of the hydraulic system used to draw water from the polar ices and pump it through to the equator.

After Lowell's death, the canals became less and less popular, though even in the 1950s Gerard de Vaucouleurs, one of the leading observers of the planet, was still claiming that they had a 'basis of reality'. It was only with the flight of Mariner 4, in 1965, that the problem was finally solved. There were no canals, in any form; they were tricks of the eye. The dark areas were not depressions, and indeed some of them were elevated plateaux; neither were they covered with vegetation, and were simply areas where the reddish, dusty material had been scorched away by Martian winds to expose the darker ground below.

The coup de grâce was given in 1976 by the two Viking probes, which were designed to make controlled landings on Mars and carry

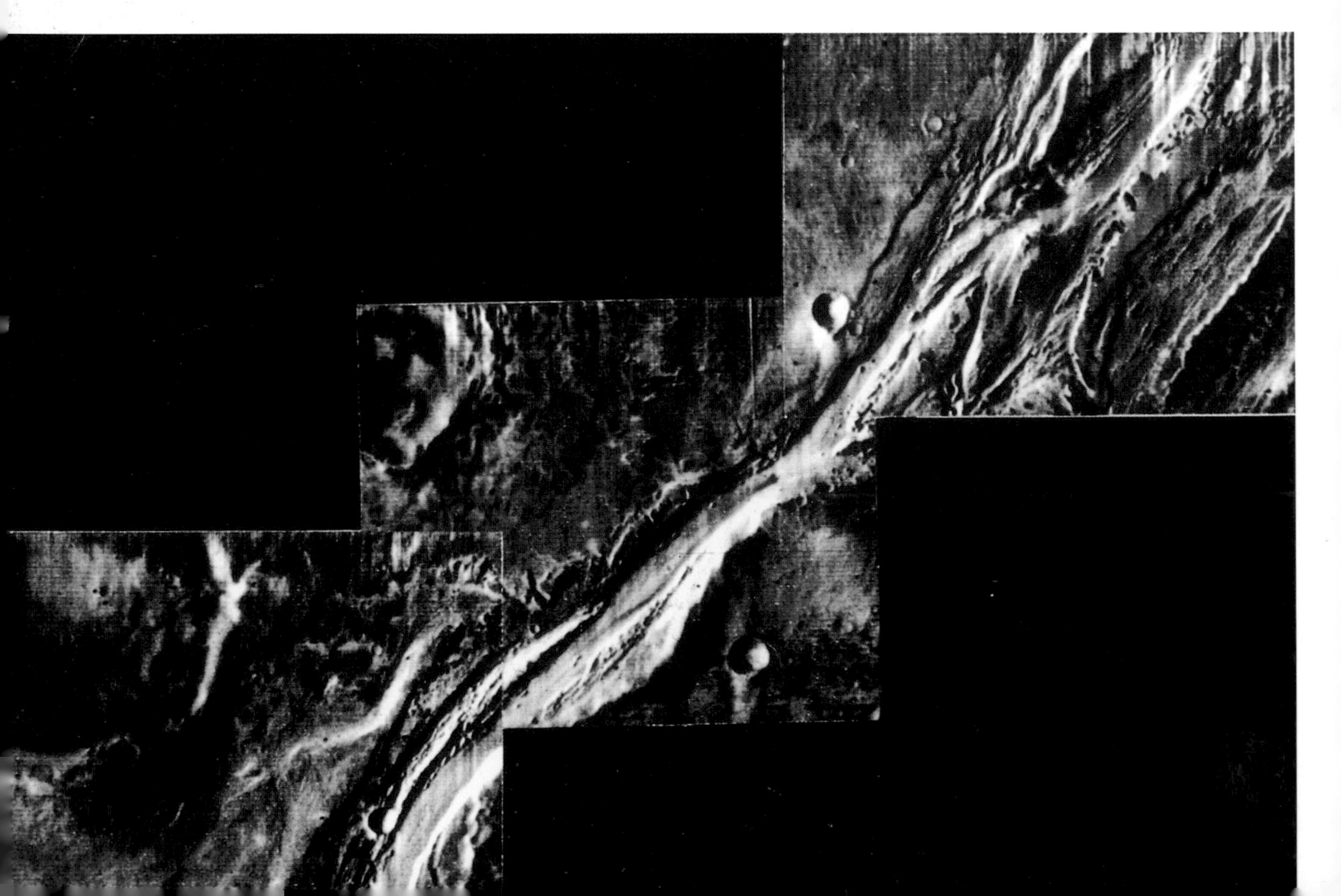

Channels on Mars from Mariner 9 (1971). It is difficult to believe that these are anything but old riverbeds, in which case Mars must once have been much more friendly than it is now

Sunset on Mars, from Viking 1

out an on-the-spot search for life. Scientists and non-scientists alike waited for the results with undisguised eagerness. Alas for our hopes! No definite trace of life was found, and it was also confirmed that the atmosphere is even thinner than had been expected, with almost no free oxygen.

Even if Mars is sterile today, as now seems highly probable, we cannot be sure that it has always been so. There is clear evidence of past running water, because the Mariner and Viking pictures show features which can only be dry riverbeds, and it is possible that Mars must have been much less infertile in the past than it is at present. There could well be striking fluctuations in climate, in which case we are seeing Mars at its very worst. We must admit, though, that even if life gained a foothold in past ages, it is unlikely to have had enough time to evolve very far. I am prepared to believe in Martian fossils of a primitive type, and when we manage to bring back samples we will know, but I am not prepared to believe in dead Martians. The first true inhabitants of the Red World will be colonists from Earth, and this may lie not so very far ahead; it is quite likely that the first Martian astronaut has already been born.

continued

1750 the French novelist Voltaire wrote a novel, *Micromégas*, also in which Mars was credited with two moons. Up to that time no telescope in the world was powerful enough to show either Phobos or Deimos, but suggestions that Swift and Voltaire were drawing upon 'ancient knowledge' were discounted when it was pointed out that since Earth had one satellite and Jupiter four, how could Mars possibly manage with less than two?

Voyaging

On 20 August 1977, the American spacecraft Voyager 2 blasted away from Cape Canaveral. It had an exciting time ahead of it. It was due to pass by the two inner giant planets, Jupiter and Saturn; then, if all went well, it might be able to go on to the outer giants, Uranus and Neptune. Its journey would take many years, and it would never return. We on Earth were saying farewell to it.

Voyager 2 proved to be an outstanding success, because it carried out its full 'Grand Tour' of the giant planets and is still sending back signals; with luck, we should maintain contact with it until about the year 2020, by which time it will have reached the boundary of that part of the Galaxy over which the Sun's influence is dominant. This is all the more remarkable because one of its main receivers failed a few months after launch, and ever since then it has been operating on its back-up system.

Obviously, Voyager 2 was unmanned. We are still far from being able to plan a piloted trip beyond the inner part of the Solar System, and even Jupiter, closest of the giants, never comes much within 400,000,000 miles of us, so that getting there is bound to take the best part of two years. In any case, there is no chance of landing on the giants, because their surfaces are gaseous — quite apart from their numerous other disadvantages, such as very low temperatures and lethal radiation zones. However, there is no reason why we should not make the trip in imagination, so I now invite you to join me on board Voyager 2 as we prepare to begin our journey.

1977: Cape Canaveral, Florida — We are ready to go. Our sister craft, Voyager 1, will follow in a few days, but will travel by a more economical route and will reach its first target, Jupiter, before we do. Voyager 1 will then go on to Saturn, beating us by about a year. During its Saturn pass it will also make a close-range survey of Titan, Saturn's largest satellite, which is know to have a thick atmosphere, and is of tremendous interest to astronomers. If Voyager 1 succeeds, then we on Voyager 2 will be able to go on to the outer planets. If not, then we will have to tackle Titan, and forget about Uranus and Neptune. We can only hope for the best.

'League of Planets'

In 1919 an amateur weather forecaster named Alberto Porta, who was born in Italy but lived in Los Angeles, announced that during December of that year all the major bodies of the Solar System would be lined up, and that this 'League of Planets' would produce 'hurricanes, lightning, colossal storms . . . there will be gigantic lava eruptions, to say nothing of floods and fearful cold'. Predictably, nothing happened, but it is on record that during the period from 17-20 December miners in Oklahoma refused to go underground in case they were trapped when the eruptions started. This is surely an original reason for a miners' strike!

Launch of Voyager 2, in 1977, bound for the outer planets

This is a good time to start, because the four outer giant planets are strung out in a long curve, and we can use their pulls to help us on our way — a procedure known officially as the gravity-assist technique, though more often termed interplanetary snooker. For example: as we approach Jupiter we will pick up speed, giving us an extra thrust out toward Saturn, while Saturn will send us on to Uranus and Uranus will help us to Neptune. We should reach Neptune in 1989; without this technique, the journey would take thirty years. The 'curve' will not recur for well over a century and a half, so that the NASA technicians have had to work hard to make the best possible use of the opportunity which Nature has so kindly provided.

Jupiter from Voyager

1978 — We have left Earth far behind, though the chance of taking a snapshot of the Earth and Moon together was really too good to be missed. Now we are far out, and Earth has shrunk into the distance; it looks like a bright, bluish star, but already it is inconveniently close to the Sun in the sky, and we must be very careful not to damage our cameras by pointing them sunward.

Callisto from Voyager: an icy, cratered, inert world

> **Near misses**
>
> In 1937 a tiny asteroid, Hermes, bypassed the Earth at only 485,000 miles, and led to the following headline in one London daily newspaper: WORLD DISASTER MISSED BY THREE HOURS AS TINY PLANET HURTLES PAST. In 1989 an even smaller asteroid, 4581 Asclepius, passed by at 370,000 miles, and then, in 1991, a real midget, 1991 BA, passed by at a mere 106,000 miles – less than half the distance of the moon.

Jupiter is still tiny in the distance. At the moment we are passing through the belt of asteroids, which are tiny worlds; only a few are as much as a hundred miles across, and these we can easily avoid, because their orbits are known. The real danger comes from much smaller particles. If we happen to collide with a piece of rock the size of, say, a writing desk, our mission will come to an abrupt end. The NASA planners have always been worried about this, but there is not much that they can do about it, and we can take heart from the fact that two earlier Jupiter probes, Pioneers 10 and 11, have already run the gamut of the asteroid zone and have emerged unscathed.

Up to now there have been no major impacts either upon our craft or upon our twin, Voyager 1, which is now just about overhauling us in the race to Jupiter. It looks, therefore, as if there are fewer small asteroids than had been feared.

January 1979 — Yes, we are through the asteroid belt, and although it is true that there are some 'strays' which swing clear of the main swarm it seems that this particular danger is over. We can look forward to Jupiter.

> **Asteroids or Planetesimals?**
>
> Recently it has been found that some bodies orbit the Sun at a surprisingly great distance. Asteroid 5145, Pholus, moves between 689 million and 2,976 million miles from the Sun, so that its orbit crosses those of Saturn, Uranus, and Neptune; the orbital period is 93 years. In 1992, D. Jewitt and J. Luu discovered 1992 QB1, which is even more extreme; the distance from the sun ranges between 3,685 million and 4,126 million miles, and the period is 196 years. Both it and Pholus seem to be around 150 miles across. Are they asteroids, or are they 'planetesimals', objects left over, so to speak, when the main planets were formed four and a half thousand million years ago?

May 1979 — The scene is breathtaking. From the messages sent back by Voyager 1, we know that we will have to keep well away from the regions where Jupiter's radiation is at its worst, as otherwise our sensitive instruments will be put out of action, but we can still skim to within 100,000 miles of the clouds, and Jupiter is spread out for our inspection. The clouds are vividly coloured, and there too is the Great Red Spot, a vast whirling storm whose surface area is as great as that of the entire Earth. Close beside it are strange white ovals. We can see the darkish belts, the loops and the festoons; our instruments can measure the winds, and Jupiter is indeed a violent place in constant turmoil. There, too, is the thin, blackish ring, quite undetectable from Earth and so very different from the glorious icy rings of Saturn.

ASTEROIDS

The asteroids, or minor planets, are very junior members of the Sun's family. Most of them move in that part of the Solar System between the orbits of Mars and Jupiter, though there are some which swing away from the main swarm. All are small; there are only four with diameters greater than 250 miles — Ceres (596 miles), Vesta (360), Pallas (325) and Hygeia (257). Only Vesta is ever visible with the naked eye.

It used to be thought that the asteroids represented the débris of an old planet or planets which met with disaster in the remote past, but it is now more generally believed that no large planet could form in this region because of the powerful pull of Jupiter; whenever a planet began to form, Jupiter literally broke it apart, so that the end product was a swarm of dwarfs.

Ceres, the largest of the asteroids, was discovered by the Italian astronomer G.Piazzi on 1 January 1801, the first day of the new century. Piazzi was not looking for a planet, and was merely drawing up a new star-catalogue when he found a star-like object which moved perceptibly from night to night. However, there had already been suggestions that a planetary body might be moving in this region of the Solar System, and a systematic search had been organised by a group of astronomers under the leadership of Johann Schröter and the Baron von Zach. Within the next few years three more asteroids were discovered — Pallas, Juno and Vesta. The fifth, Astræa, was found in 1845, and today the total known membership of the swarm is over 3,000. In all, there may be at least 40,000 asteroids, though most of them are very tiny, and are irregular in shape. As yet nothing is known about their surface features, because even in giant telescopes they look exactly like stars. We must await the first fly-by of an asteroid by a spacecraft, which will almost certainly happen in the near future.

Perhaps the most interesting asteroids are those which depart from the main zone, and may come close to the Earth. One of these is Eros, first seen in 1898, which is shaped rather like a sausage, with a longest diameter of 18 miles; it can approach us within 15,000,000 miles, as happened in 1931. Other 'Earth-grazers' can come even closer, and a few have scraped past us at less than twice the distance of the Moon, but all these are small by any standards. Hathor, which made a close approach in 1976, has a diameter of no more than 352 yards!

Other asteroids, the Trojans, lie well beyond the main zone, and share the same orbit as Jupiter, though they keep prudently either well ahead of or well behind the Giant Planet, and are in no danger of being swallowed up. Some small asteroids have very eccentric orbits; Hidalgo, for example, has a path which takes it from near the orbit of Venus out almost as far as Saturn. And there are two asteroids, Icarus and Phæthon, which invade the torrid regions within the orbit of Mercury, so that when at perihelion they must be red-hot.

One of the oddest of the asteroids is Chiron, discovered in 1977. It spends most of its time between the orbits of Saturn and Uranus, and has a revolution period of fifty years; the estimated diameter is of the order of 150 miles. It is now coming in toward its perihelion, and has been seen to develop a 'haze', so that there have been suggestions that it could be a comet rather than an asteroid. Against this, Chiron seems much too large to be a comet, and more probably it has an icy surface layer which begins to evaporate when near perihelion, producing a temporary 'atmosphere'. At present Chiron is very much of an enigma, and its exact nature is very uncertain.

Closing in on Saturn: 8,600,000 miles on 11 August 1981 (Voyager)

We can also see that Jupiter is spinning quickly — its 'day' is less than ten hours long — and that the clouds shift and change rapidly. The belts are made up of droplets of ammonia, and below are other layers, including (we think) clouds of water ice. Unfortunately we cannot see them, and we will not come this way again.

But wait! Jupiter is not alone. It has a whole family of satellites or moons, and by now we can see that the four large members of the system are quite unlike each other. Look first at Callisto and Ganymede, the outer two, both of which are larger than our Moon and are cratered, ice-coated and dead, so that nothing has happened upon them for thousands of millions of years. Closer in to Jupiter we can make out the smooth face of the next satellite, Europa, and we regret that we cannot go nearer, because the view is so strange. There are no craters on Europa; neither are there any mountains or valleys — nothing but a maze of shallow depressions, probably no more than a few feet deep. Europa looks almost like a cracked eggshell.

If Europa is covered with a layer of ice, there may well be liquid water underneath, where the temperature is less bitter. There have even been suggestions that there may be primitive life in this sunless sea. It may sound improbable, but we cannot be sure, and to make a closer study of Europa we must wait for a new mission. Already we are past, and on our way to an encounter with the innermost of Jupiter's large moons, Io.

Surprise follows surprise. There is nothing icy or inert about Io. It is red, and looks remarkably like an Italian pizza; here and there we can make out dark patches, and we can see that there are active volcanoes, sending material high above the ground. Multi-coloured débris must rain down on the surface, and we find that the redness is caused by sulphur, so that the volcanoes are very unlike those of Earth. To complete the picture, we find that Io is moving right in the midst of Jupiter's deadly radiation. It has been said that Venus lays claim to being the most hostile world in the Solar System, but from our present vantage point we think that the palm must be handed to Io. We will go no closer; if we want to continue our surveys, we must leave as soon as possible.

Destroy or divert

Calculations have been made by a group at the Massachusets Institute of Technology to see whether it would be possible to destroy or divert an asteroid found to be on a collision course. One suggestion is to blow it to bits; another idea involves landing a nuclear bomb on it and then causing an explosion violent enough to throw the asteroid into a new orbit.

Saturn's rings

The rings of Saturn were first seen by Galileo, in 1610, with his primitive telescope, but he could not make out what they were, and thought that Saturn must be a triple planet. When he looked again in 1612 he was astonished to find that the two smaller bodies had disappeared, and he wrote: 'Have they vanished or suddenly fled? Has Saturn, perhaps, devoured his own children?' We now know that in 1612 the rings were tilted edgewise-on to the Earth, so that no small telescope could show them.

The true nature of Saturn's rings was discovered in 1656 by the Dutch observer Christiaan Huygens. Sir Christopher Wren, at that time professor of astronomy at Oxford, had had a different idea, but at once abandoned it in favour of Huygens' explanation, because, as he said, 'I loved it better than my own'.

JUPITER

Jupiter is much the largest of the planets, and is more massive than all the others put together. Despite its distance it is a brilliant object in our skies; it is outshone only by Venus and, very occasionally, by Mars.

Jupiter is quite unlike the inner planets. Telescopically it appears as a yellowish, flattened disk, crossed by dark 'belts' and with various spots, wisps and festoons. The surface is gaseous, and is always changing. The globe is flattened because of the rapid rotation, making the equator bulge out; though the Jovian 'year' is almost twelve times the length of ours, the 'day' is less than ten hours long. Moreover, Jupiter does not rotate in the way that a rigid body would do. The equatorial zone spins round in 9hr 51min, while the rest of the planet takes five minutes longer, but special features such as spots have periods of their own, so that they drift around in longitude.

It used to be thought that Jupiter must be a miniature star, sending out enough energy to warm its system of satellites, but this is not true; though the core has a temperature of at least 30,000°C, the outer layers are intensely cold. According to modern theory there is a relatively small, iron-rich core, surrounded by deep layers of liquid hydrogen; above these layers comes the 'atmosphere', several hundreds of miles in depth. Hydrogen makes up most of it, together with hydrogen compounds such as ammonia and methane together with a considerable quantity of the second lightest gas, helium. This is no surprise; hydrogen and helium are much the most plentiful elements in the entire universe.

Gases warmed by Jupiter's internal heat rise into the high atmosphere and cool, forming clouds. Apparently there are various cloud-layers of different composition, one of which is made up of crystals of water ice, while the uppermost clouds are due to ammonia crystals. The dark belts are regions where gases are descending.

The most famous marking on Jupiter is the Great Red Spot, which was first reported during the seventeenth century and has been on view for most of the time ever since, though it sometimes disappears briefly. Telescopes show it as a prominent reddish oval; its surface area is greater than that of the Earth, and its maximum length is not far short of 30,000 miles. Before the Space Age it was often believed to be a solid body floating in Jupiter's outer gas, but we now know it to be a vast whirling storm — a phenomenon of Jovian 'weather'. The reason for its colour is not definitely known, but could be due to phosphorus. There are many other spots, but none with the same striking hue, and in any case most of them are short-lived. Jupiter is an active world, with a surface which is in a constant state of turmoil.

Four spacecraft have so far bypassed Jupiter: Pioneer 10 (1973), Pioneer 11 (1974) and Voyagers 1 and 2 (1979). They have found that the planet is associated with zones of radiation which would be lethal to any astronaut foolish enough to venture inside them, and that there is a very strong magnetic field; it had already been established that Jupiter is a source of radio waves. The Voyagers recorded a thin, dark ring, which is too faint to be seen from Earth and is quite unlike the glorious icy rings of Saturn.

Jupiter has a whole retinue of satellites. Four of them (Io, Europa, Ganymede and Callisto) are large, and were discovered by Galileo as long ago as 1610, so that they are known collectively as the Galileans. All the other satellites are less than 200 miles in diameter. Metis, Adrastea, Amalthea and Thebe move closer in to Jupiter than the orbits of the Galileans; Leda, Himalia, Lysithea, Elara, Ananke, Carme, Pasiphaë and Sinope are further out. The last four move round Jupiter in a wrong-way or retrograde direction, so that they are probably captured asteroids. Data for the Galileans are as follows:

Name	**Mean distance from Jupiter (miles)**	**Revolution period (days)**	**Diameter (miles)**
Io	262,000	1.769	2,264
Europa	417,000	3.551	1,945
Ganymede	666,000	7.155	3,274
Callisto	1,170,000	16.689	2,987

All were surveyed by the Voyagers, and have proved to be remarkable objects, particularly the highly volcanic Io. A small telescope will show them as starlike points, and it is fascinating to watch them from night to night as they move round Jupiter; they may pass in transit across the planet's face, they may be occulted, or they may be eclipsed by Jupiter's shadow. We should learn even more about them in 1995, when the latest Jupiter probe, named in honour of Galileo, reaches its target.

1980 — This is the period of 'cruise mode'. We can still make measurements which are of great value to our controllers on Earth, now well over 600,000,000 miles away, but we are out of the range of any planet; Jupiter is far behind, Saturn far ahead. Yet we know that Voyager 1 has completed its survey of Titan as well as Saturn, so that our own way is clear for Uranus and Neptune.

1981 — If Jupiter is the most majestic of the planets, then Saturn is without doubt the loveliest. Even from Earth it is a superb sight, and now, as we on Voyager 2 approach it, we can see that it is even more striking than we had expected. True, the surface is not so vividly coloured or as active as that of Jupiter, and the overlying 'haze' gives Saturn a relatively bland appearance, but the rings defy description.

For many years Earth-based astronomers have known of three

False colour view of Saturn's surface from Voyager 2. The colours indicate differences in composition, but do not represent what would actually have been seen by an observer on the space-craft

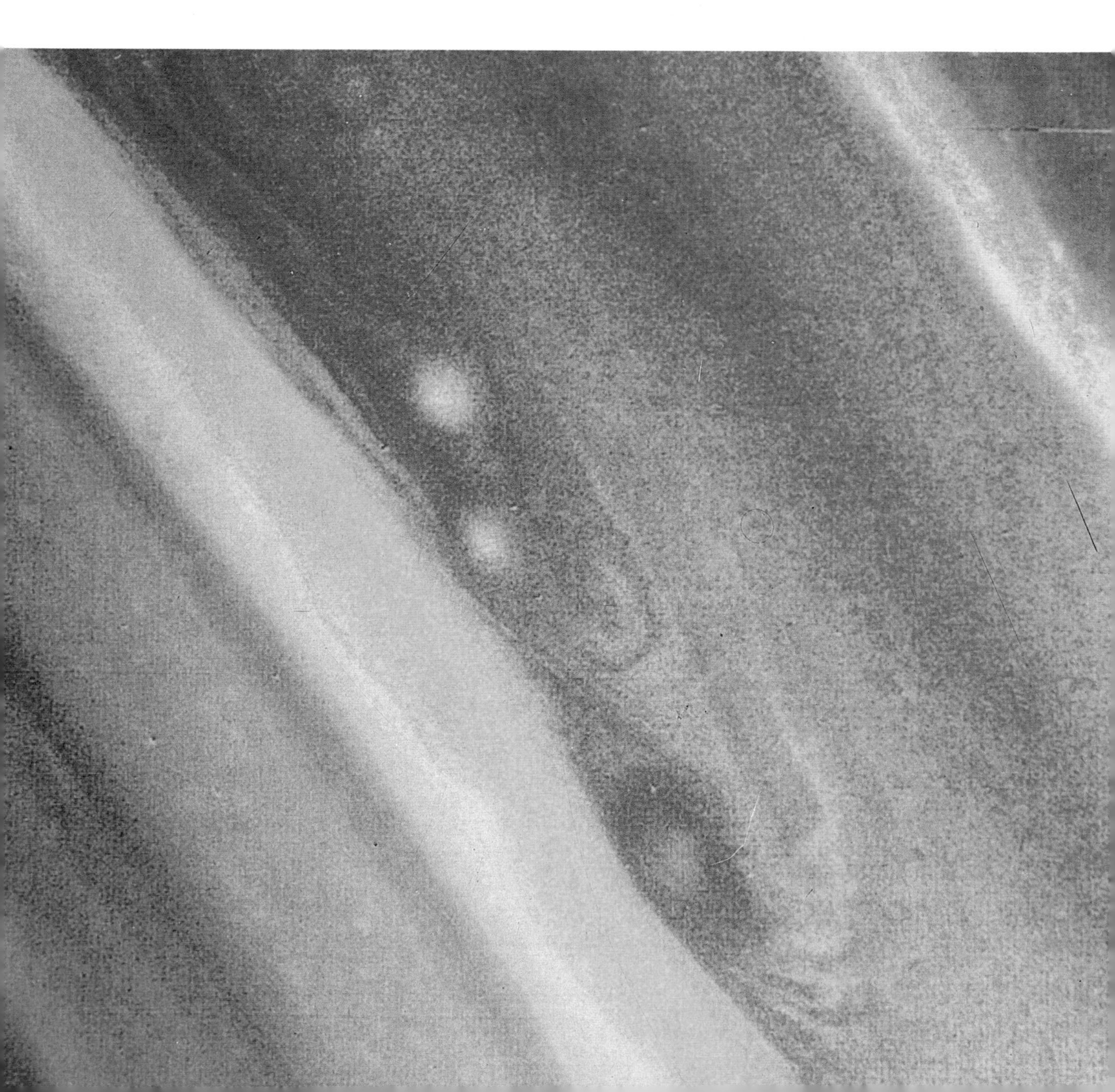

SATURN

Saturn was the outermost of the planets known in ancient times. With the naked eye it looks like a brightish but rather dull, slightly yellowish star, and the old astrologers believed that it had a baleful influence, but telescopes show it to be without doubt the most beautiful object in the entire sky.

Saturn is second in size only to Jupiter, and is basically the same kind of world. There is a rocky core, at a temperature of around 15,000°C, overlaid by layers of liquid hydrogen, above which comes the deep 'atmosphere' rich in hydrogen and hydrogen compounds such as methane, together with helium. As with Jupiter, the upper clouds are very cold indeed. The rotation period is less than eleven hours, so that the disk is very obviously flattened.

Belts on Saturn are much less conspicuous than those of Jupiter, and there is more 'haze', giving the planet a somewhat bland appearance. Well-defined spots are rare, and there is nothing comparable with the Great Red Spot on Jupiter. In 1933 W.T. Hay discovered a bright white spot near the planet's equator, but it did not last for long. Another prominent white spot appeared in 1990.

Of course the main glory of Saturn is its system of rings, which can be well seen with even a small telescope when they are suitably tilted toward us. They look solid, but no solid or liquid ring could possibly exist, because it would at once be pulled apart by the strong gravitational pull (and remember, Saturn is ninety-five times as massive as the Earth). The rings are made up of vast numbers of icy particles, ranging in size from 'pebbles' up to objects the size of a writing desk, all whirling round the planet in the manner of dwarf moons. From end to end the system measures 170,000 miles, but it is no more than a mile thick at most, so that when the rings are placed edgewise-on to us — as happens on average every fifteen years or so — they practically disappear.

There are two main rings, separated by a 2,500-mile gap known as Cassini's Division in honour of its discoverer (G.D. Cassini, in 1675). An inner, semi-transparent ring was found in 1848, and is generally known as the Crêpe or Dusky Ring.

Saturn has so far been bypassed by three probes, Pioneer 11 (1979), Voyager 1 (1980) and Voyager 2 (1981). The Pioneer encounter was more or less an afterthought by the planners, as Jupiter had been the main target, so that most of our knowledge of Saturn comes from the Voyagers, both of which were outstandingly successful. The rings proved to be very complex, with thousands of narrow divisions; there were even ringlets in the Cassini Division, and two outer rings were found, one of which is curiously 'braided'. The Voyagers also sent back information about Saturn's magnetic field, which is very marked even though much weaker than that of Jupiter. The radiation zones, too, are less intense.

Saturn has eighteen known satellites, of which the first and largest, Titan, is visible with a small telescope, and has been known ever since 1655. Data for the satellites with diameters of over 120 miles are as follows:

Name	Mean distance from Saturn (miles)	Revolution period (days)	Diameter (miles)
Mimas	115,300	0.941	247
Enceladus	147,950	1.370	310
Tethys	183,140	1.888	650
Dione	234,570	2.737	696
Rhea	327,560	4.518	950
Titan	759,390	15.495	3,201
Hyperion	920,510	21.277	223x174x140
Iapetus	2,213,360	79.331	892
Phœbe	8,050,960	550.4	137

Of the rest, Atlas, Prometheus, Pandora, Epimetheus and Janus are closer in than Mimas; Telesto and Calypso move in the same orbit as Tethys, while Helene shares the orbit of Dione. Phœbe has retrograde motion, and is probably an ex-asteroid.

Titan, the only satellite in the Solar System to have a dense atmosphere, is of special interest. If all goes well, a new probe ('Cassini') will be launched in 1995, and will reach the neighbourhood of Saturn in 2002, with Titan as its main target. The plan is to survey Titan by radar, and then send a 'lander' down through the atmosphere by parachute — though whether it will come down on solid ground or in a chemical ocean remains to be seen. All the ingredients for living matter exist on Titan, but it seems likely that the intense cold has prevented life from appearing there.

rings, two bright and one dusky and semi-transparent. Certainly there is no chance of their being solid or liquid sheets; any substantial ring would be at once pulled apart by Saturn's strong gravity (and remember, Saturn is almost a hundred times as massive as the Earth). The rings are made up of icy particles, all whirling round in the manner of dwarf moons. The main division between the two bright rings, named Cassini's Division in honour of its seventeenth-century discoverer, was once believed to be empty. Voyager 1 had shown that this is not so, and now that we are within range, we can see for ourselves that the Division contains ringlets.

The ring system itself is incredible. There are thousands of narrow ringlets and minor divisions, and in the brightest of the rings there are dark 'spokes' which ought not to be possible. Why are the rings like this? Can the appearance be due to the gravitational pulls of Saturn's satellites? A new ring, beyond the main system, is 'twisted', and seems to be kept in place by two small satellites, one just closer in to Saturn and the other just further out, which act as 'shepherds', and force any errant particle back into its proper place.

We will leave these problems to the theorists, because there is so much to see – and very little time to see it, because Saturn will be by-passed quite quickly. The satellites are of different types. In the distance we can see Titan, which was so important to the NASA planners, but there is no surface detail, and all we can see is the top of an orange layer of what we might well call smog. Voyager 1 has already found that Titan has a dense atmosphere made up chiefly of nitrogen, and there may be oceans on its surface, filled not with water but with a much less pleasant liquid called ethane. However, Titan is not our business; we are better occupied in surveying some of the other moons such as Mimas, with one huge crater; Enceladus, partly pitted and partly ice-smooth; Tethys, almost pure ice; cratered Dione and Rhea, and Hyperion, which is shaped rather like a hamburger. It is a pity that we do not go closer to Iapetus, the outermost of the large members of the Saturnian family, and so far out that it takes seventy-nine days to complete one orbit. Iapetus is almost a thousand miles across, and has a strange surface, part of which is as white as ice while another part is blacker than a blackboard. It is tempting to think of the 'zebra problem'; is a zebra a white animal with black stripes, or a black animal with white stripes? In the case of Iapetus, we can find out that it is not much denser than water, so that it must be icy. The black deposit has presumably welled up from inside the globe, though we do not know why.

Saturn is now behind us, but we have a problem. The scan platform carrying our cameras has stuck. Unless the cameras can be pointed in the right direction, there can be no pictures. We can imagine the consternation at NASA headquarters in California: what has gone wrong? Could it be that the platform has been hit by an icy ring-particle? If so, there is not a great deal to be done about it, and the Uranus-Neptune part of our mission will be in jeopardy.

Luckily, we on Voyager know the answer, even though NASA does not. It is a question of lubrication, because the controllers have

Mission to Jupiter

On 18 October 1989 the 'Galileo' probe was launched on a mission to Jupiter. To get there it has to follow a complicated path, taking it first past Venus and then making two further passes of the Earth before embarking on the final leg of its journey – rather like going from Brighton to Bognor Regis by way of Carlisle. (The reason for this is that the probe had to be launched from a rocket less powerful than the one originally planned.) Because it carries a small quantity of highly radioactive plutonium to power its equipment, there was considerable alarm in the region of the Cape Canaveral launching ground in Florida, and an organisation calling itself the Florida Coalition for Peace and Justice tried to stop the entire programme on legal grounds. This led to an interesting exchange between the chief protestor, a Mr Gagnon, and the presiding judge. Mr Rogers, leader of the protesters, said that NASA had no experience of this kind of launch (though in fact it had plenty). Judge Gaskin replied that Columbus didn't have much experience either. Mr Rogers came back with the retort that Columbus wasn't going to pollute the world.

Judge Gaskin ruled that the risk was so slight that it could be ignored. So the protesters lost their case; students from a nearby university did not carry out their threat to sit on the launch-pad, and Galileo departed for Jupiter on schedule.

Unfortunately its high-gain antenna has so far (1993) failed to unfurl, and this will reduce the amount of data received; all the same, Galileo should produce some remarkably interesting results in 1995.

WILLIAM HERSCHEL

In March 1781 an amateur astronomer named William Herschel, living in Bath, was carrying out a systematic 'review of the heavens' with a reflecting telescope which he had made himself. Suddenly he came across an object in the constellation of Gemini, the Twins, which did not look like a star. It showed a small disk, and Herschel naturally believed it to be a comet. Over the next night or two it moved, so that clearly it was a true member of the Solar System, but not for some time was it realised that it was a new planet, moving far beyond the orbit of Saturn. After some discussion it was named Uranus, after the first ruler of Olympus.

Friedrich Wilhelm Herschel was Hanoverian by birth (at that time England and Hanover were under the same crown), and came from a musical family. His father was a bandmaster, and young Wilhelm joined the Hanoverian Army, but after an uncomfortable experience during the wars with France he decided that a military career was not for him, so he left the Army and came to England. It is not correct to say that he deserted; he had been too young to be formally enlisted, so that he was quite within his rights in leaving. He spent the rest of his life in England, and we always know him as William rather than Wilhelm.

He embarked upon a musical career, and became an organist in the fashionable centre of Bath, where he soon built up a considerable reputation. During the 1770s he became seriously interested in astronomy; after many failures, and with the invaluable help of his sister Caroline he made his first good telescope, and then started an observational programme which lasted almost up to the time of his death in 1822. His discovery of Uranus made him world-famous; he was appointed Court Astronomer to King George III of England and Hanover, and was able to give up music as a career, though he always retained his keen interest in it. After leaving Bath he settled in Slough, and it was here that he set up his largest telescope, with a mirror 49in in diameter — easily the largest ever made up to that time.

Herschel became the best observer of his day, and discovered thousands of double stars, clusters and nebulæ as well as being the first man to draw up a reasonably good picture of the shape of our star-system or Galaxy, but he will always be best remembered for his discovery of Uranus — the first new planet to be found since ancient times.

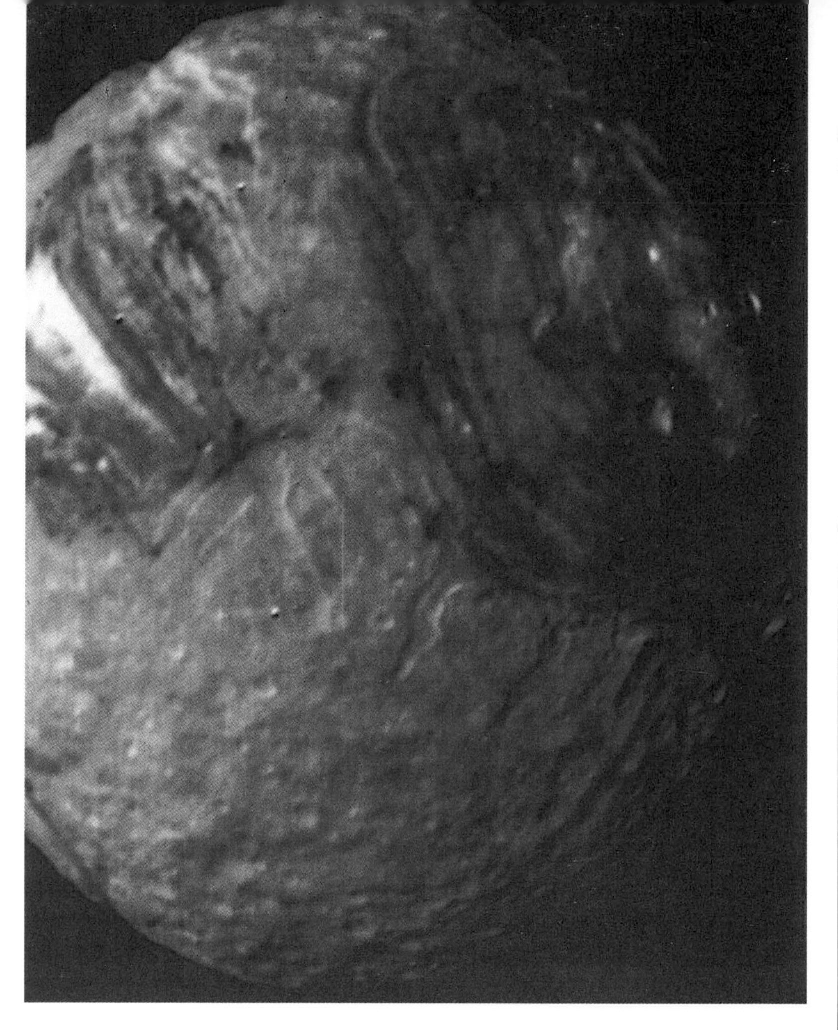

Miranda, with its immensely varied surface, taken from Voyager

been slewing the platform round too quickly, and it has seized up. Regrettably, there must be some loss of data as we move away from Saturn, but if the ground technicians realise what has happened they may be able to put it right.

1982 — All is well. Operating the scan platform at a slower rate means that it can still be accurately aimed. We are back in cruise mode, and we hear that at NASA the technicians are busily updating their software, so that when we reach Uranus in four years from now the pictures received on Earth should be every bit as good as those from Saturn, even though the light-level at Uranus is so much lower.

January 1985 — Still in cruise mode. It is easy to forget how remote Uranus is; more than twice as far as Saturn. It still looks like a tiny disk, and in any case we cannot hope to see as much detail as with the inner giants, partly because Uranus is smaller and no doubt less active, but also because we will have to approach it pole-on; the tilt of the axis is more than a right angle, and at present the south pole is facing the Sun. Uranus has long seasons, as it takes eighty-four Earth years to complete one orbit.

December 1985 — Now, at last, we are within striking distance of

Mirror mishap

When William Herschel discovered Uranus he was observing from the garden of his house, 19 New King Street, Bath, now preserved as a Herschel museum. He used to make mirrors in the cellar, and on one occasion there was a slight mishap, vividly described by his sister and invaluable helper Caroline:

> The mirror was to be cast in a mould prepared from horse dung. . . . By 11 August 1781 all was ready, but there was an initial failure. The molten metal ran into the mould easily enough, but the mould itself began to leak . . . the result was disaster. The stream of hot metal grew from a trickle into a torrent and spread across the floor. Both my brother and the caster and his men were obliged to run out at opposite doors, for the stone flooring (which ought to have been taken up) flew about in all directions as high as the ceiling. My poor brother fell exhausted by heat and exhaustion on a heap of brickbats.

Opposite:
The rings of Uranus from Voyager 2. Unlike the rings of Saturn, Uranus' rings are thin and narrow

URANUS

Uranus is just visible with the naked eye if you know where to look for it, but it is not surprising that nobody had identified it as a planet before Herschel did so in 1781. It has less than half the diameter of Saturn, but it still ranks as a giant, and its mass is fourteen times that of the Earth.

Perhaps the most curious fact about Uranus is that its axis of rotation is tilted at an angle of 98° — more than a right angle, so that the 'seasons' there are very unfamiliar. First one pole, then the other, has a 'midnight sun' lasting for twenty-one Earth years, with corresponding darkness at the opposite pole; for the rest of the long 'Uranian' year conditions are less extreme. Since the planet spins once in 17¼ hours, there are over 42,000 'days' in each 'year', and at times the equator is the coldest part of the planet. The reason for this strange tilt is unknown. There have been suggestions that in its early history Uranus was hit by a massive object and literally knocked on to its side; this does not sound very plausible, but it is not easy to think of anything better.

Unlike the other giant planets, Uranus has no strong source of internal heat. There is presumably a rocky core; this is surrounded by a dense layer in which gases are mixed with 'ices', ie substances which would be frozen at a lower temperature. These 'ices' are mainly water, ammonia and methane, which condense in that order to form thick, icy cloud layers. Above comes the 'atmosphere', made up chiefly of hydrogen together with a considerable quantity of helium and lesser amounts of other gases.

Little can be seen from Earth on the pale, green disk of Uranus, and there are no spots or clouds like those of Jupiter or Saturn. However, in 1977 it was found that there is a system of thin, dark rings. They were discovered by chance, when astronomers were observing an occultation — that is to say, the disappearance of a star behind Uranus as the planet passed in front of it. Both before and after the actual occultation the star 'winked' regularly, showing that it was being briefly hidden by rings surrounding the planet. Unlike the rings of Saturn, they are blacker than coal-dust, and most of our knowledge of them has been sent to us by Voyager 2.

The Voyager pass took place in January 1986. Only a few clouds were seen, but the rings were studied in detail; they seem to be made up of particles a few feet in diameter, and to contain plenty of 'dust'. A magnetic field was detected, and it was confirmed that Uranus is a source of radio waves.

Five satellites — Miranda, Ariel, Umbriel, Titania and Oberon — were known before the Voyager mission. Voyager discovered ten more, all closer in than the orbit of Miranda. They have been given Shakespearean names: Cordelia, Ophelia, Bianca, Cressida, Desdemona, Juliet, Portia, Rosalind, Belinda and Puck. Puck, with a diameter of 96 miles, is much the largest of them. Data for the other satellites are as follows:

Name	Mean distance from Uranus (miles)	Revolution period(days)	Diameter (miles)
Miranda	81,100	1.414	293
Ariel	119,200	2.520	720
Umbriel	166,070	4.144	727
Titania	272,220	8.706	981
Oberon	364,300	13.463	947

Herschel discovered Titania and Oberon in 1787, with his 49in telescope, and may have glimpsed Umbriel. Umbriel and Ariel were definitely seen by the English amateur William Lassell in 1851, and Miranda was discovered by G.P. Kuiper in 1948.

We cannot claim that our knowledge of Uranus and its system is at all complete, but to learn more we must wait for the dispatch of a new spacecraft to the planet.

Uranus. Nothing at all can be seen on the pale, greenish disk, but at least we can make out the narrow ring-system, and our cameras have just discovered a new satellite to add to the five already known. The NASA planners have even arranged to take a picture of it as we fly by.

26 January, 1986 — Everything is happening with frightening speed! We are flashing toward Uranus, and within twenty-four hours we will have collected all the close-range data we can expect. Already we have found ten new inner satellites, including the discovery made last month, and a few extra rings, but what about the magnetic field, and

any possible radio emissions? As yet they have not been detected, and neither are there any definite clouds in the upper atmosphere.

Yet as we move in, we do at last pick up radio waves, and there is a fairly powerful magnetic field. Another surprise here; the magnetic axis must be tilted to the axis of rotation by a full 60°, and is also offset from the centre of the globe for reasons which we cannot understand. Now for the clouds — they are only just detectable, very different from the frenzied activity on Jupiter or even Saturn. This could well be due to the fact that Uranus, unlike the other giants, has no marked source of internal heat.

There is a strange luminosity over the day side of the planet, which we call the electroglow, but it is time to turn our attention toward the five larger satellites. Oberon and Titania are icy and cratered, as might have been expected. Umbriel, rather smaller, has a darkish surface, but near the edge of the disk there is one bright feature which might be a crater, though it is so foreshortened that we cannot be sure. Ariel, about the same size as Umbriel, has ice, craters, and strange branching valleys which look as though they had been cut by water, even though Ariel is much too insignificant to hold on to any trace of atmosphere, and it is not easy to see how liquid water can ever have flowed there.

Looking back at Uranus' rings, we can see an unexpectedly large quantity of 'dust' being lit up by the Sun's rays, but there is no chance to linger; we are about to skim over Miranda, the smallest of the Uranian satellites known before our cameras picked up the inner swarm. Miranda is a mere 293 miles in diameter, but already we can see that it is unusual. Now, as we swing close, we can make out an amazingly varied landscape. Parts of it are old and cratered, and to one side there is an ice cliff almost five miles high, but what are those peculiar enclosures which look almost like race-tracks? Geologists will puzzle for a long time over the pictures of Miranda. It may be that the little satellite was broken up early in its history, only to re-form later. It is unlike anything we have so far seen in the Solar System.

The poles of Uranus

Because Uranus has an axial tilt of more than a right angle, it is not easy to decide which pole is which. The International Astronomical Union has decreed that all poles above the ecliptic (*ie* the plane in which the Earth orbits) are north poles, and all poles below the ecliptic are south poles. In this case, the pole of Uranus which was in sunlight at the time of the Voyager 2 pass was the south pole. However, the Voyager team members had a different system. If you grasp the spin axis of a ball with your right hand so that your fingers curl round in the direction of the ball's rotation, then the north pole is the one in the direction of your thumb. Following this rule, the sunlit pole during the Voyager mission was the *north* pole. Take your pick!

1987 — The Uranus encounter seems long ago, and once more we are back in our cruise mode. Earlier on, the NASA technicians had suggested that there was a 60 per cent chance of our being able to send back useful data from Uranus, but only a 40 per cent chance from Neptune. We hope to prove them wrong. The updated software at Mission Control means that our on-board equipment is functioning even more effectively than it did after launch, a full decade ago.

1988 — Still on schedule, and no major course corrections are needed. The first far-encounter pictures of Neptune have been sent back. They are rather smeared, because no attempt was made to correct for our own motion relative to Neptune, but already they show the point of light which marks Triton, the senior of Neptune's two known satellites. Some astronomers believe it to be larger than our Moon, and to have a dense, cloudy atmosphere. Before long we will find out whether or not this is true.

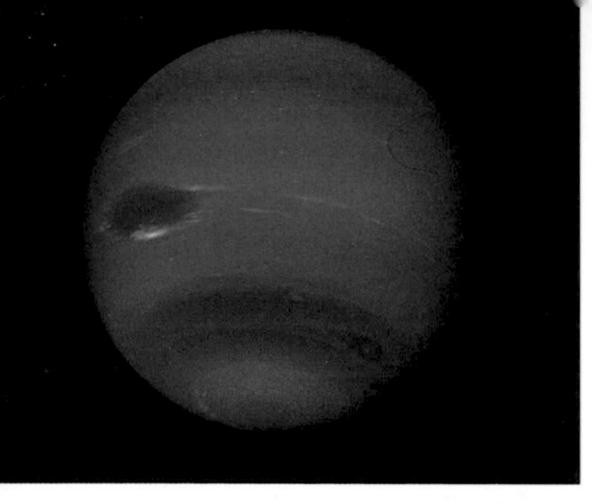

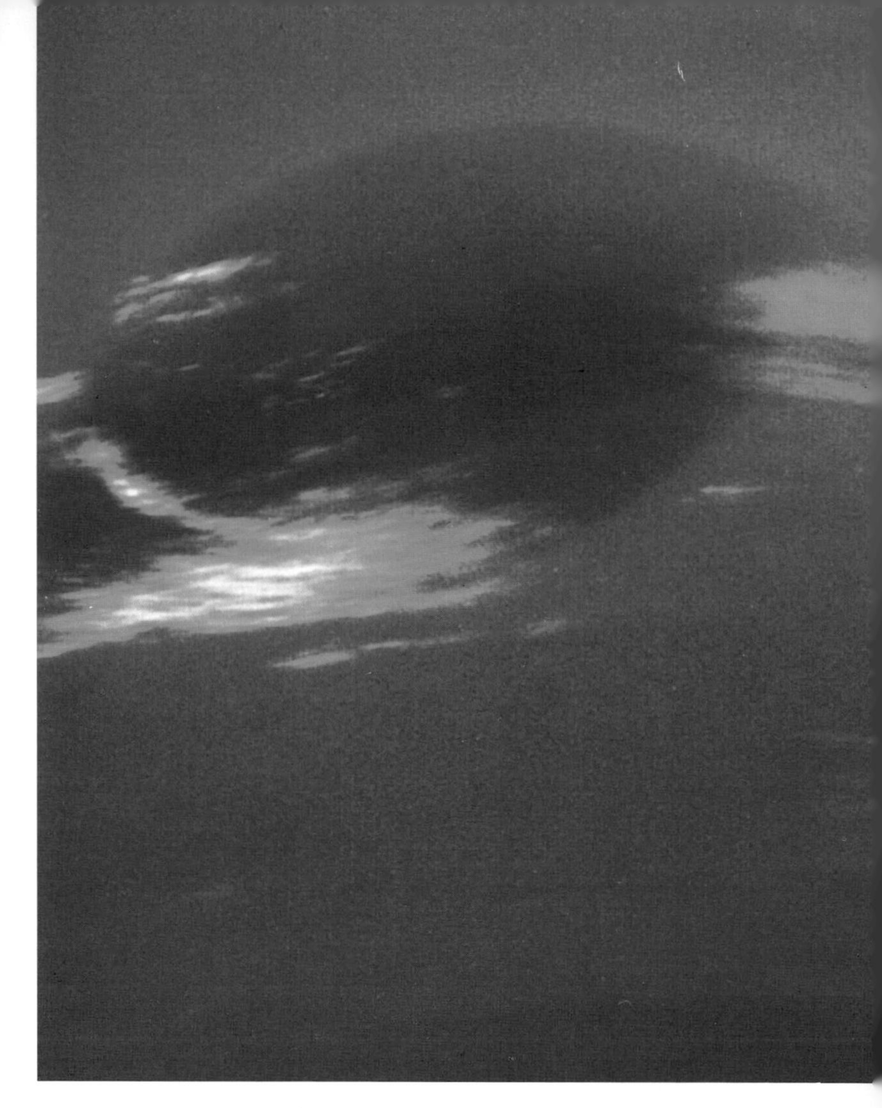

Above:
Neptune, from Voyager 2, with its beautiful blue disk and the Great Dark Spot
and right:
Closing in on the Great Dark Spot

We will also be able to decide whether or not Neptune has rings. There seems no valid reason why not — after all, Saturn and Uranus have extensive systems, and even Jupiter is encircled by a thin, dark ring — but astronomers have never been able to be sure; they have suspected 'ring arcs', or incomplete sections of rings, but that is all. If we on Voyager 2 cannot find out, then astronomers on Earth will have to wait for many decades before knowing. Remember, the gravity-assist technique cannot be used again until well into the twenty-second century, at least insofar as Neptune is concerned.

10 August 1989 — Neptune has already come up to expectations. From our vantage point we can see a gloriously blue disk, on which there is at least one large, dark feature together with obvious whitish clouds. It is summer in Neptune's southern hemisphere; we will skim only three thousand miles above the darkened north pole. Out here the sunlight is painfully weak, but the improvements in NASA's ground equipment more than compensate for this. We are ready for Encounter Day.

26 August 1989 — Neptune is amazing. We have already sent back

pictures of the ghostly ring system, which is not far above our threshold of visibility; the main ring is 'clumpy', which misled Earth observers into believing that the rings must be incomplete. Six new inner satellites have been found, Proteus, Larissa, Galatea, Despina, Thalassa and Naiad – and we have managed to send back close-range pictures of two of them, but Neptune itself dominates the scene. On its surface is the Great Dark Spot, a huge storm above which hang ice-crystal clouds of methane, while closer to the south pole there is a second spot with a bright centre, and there is also a faster-moving feature which has already been nicknamed the Scooter because of its quick rotation. Winds on Neptune are strong, reaching well over 700mph in some latitudes, and the magnetic field, like that of Uranus, is tilted and offset. There are radiation zones, though they do not seem to be strong enough to damage our equipment. Neptune is a dynamic world, totally unlike its docile twin Uranus.

We would like to pause, but we cannot, and there is no turning back. We must pass through the ring-plane, and for a few minutes there is a danger of damage, but we are soon through, and on course for our final target. Triton is now coming into view, and is clearly going to be the highlight of the whole mission.

Order and method

Sir George Airy had a great passion for order and method. He once spent a whole day in the cellar of Greenwich Observatory, marking empty boxes 'Empty'. He insisted that his observers should be on night duty even when it was raining, and he used to go round in the early hours tapping on the domes and saying 'You are there, aren't you?' It is said that his ghost still does so today, though I have never encountered it myself.

Frozen lake on Triton. This is away from the southern cap, in the northern area which has been enduring its long 'winter' (Voyager 2)

It is smaller than we had thought, with a diameter much less than that of the Moon. If it is smaller than expected, it must also be more reflective, and therefore colder; we can measure the temperature at about -400°F, so that Triton is easily the coldest world ever encountered by a space-craft. We need not worry about all-concealing clouds, because the atmosphere is very thin, and seems to be made up chiefly of nitrogen. There is some haze above the

THE DISCOVERY OF NEPTUNE

Neptune has a unique place in the history of astronomy. It was 'discovered' before it was actually seen!

As soon as Uranus had been identified, astronomers set to work in calculating its orbit. They soon found that something was wrong, because Uranus refused to behave; it persistently wandered away from its predicted path. Two mathematicians, John Couch Adams in England and Urbain Le Verrier in France, independently decided that there must be a disturbing influence, presumably a planet moving at a much greater distance from the Sun. They were faced with a sort of cosmical detective problem; they could see what was happening to Uranus, and they had to track down the culprit. Quite unknown to each other, they calculated a position for the new planet — and then came a strange chapter of accidents which has caused controversy ever since.

Adams finished first, in 1845, and sent his results to the Astronomer Royal at Greenwich, Sir George Airy. Unfortunately Airy was not inclined to take much notice of the work of a young, unknown mathematician, and nothing was done. Adams was discouraged, and failed to follow the matter up.

Le Verrier finished his work some time later, but his colleagues in France were as uncooperative as Airy had been, and Le Verrier sent his results to the Berlin Observatory, asking the Director, Johann Encke, to begin a search. Two of Encke's assistants, Johann Galle and Heinrich D'Arrest, took up the challenge — and on the very first night of their hunt they found a starlike object which proved to be the planet we now call Neptune. It was very close to the position which Le Verrier had given.

Meanwhile, news of these developments had reached Airy, and, belatedly, he realised that something would have to be done. There was no suitable telescope at Greenwich, so Airy instructed the professor of astronomy at Cambridge, James Challis, to search for the planet with the powerful refractor there. Challis was not particularly enthusiastic; unlike Galle and D'Arrest, he had no good star maps of the area, and he plodded on, checking each star visually. Actually he saw Neptune very early in his search, but did not recognise it as a planet until he heard the news from Berlin.

When the French found that Adams had come to the same conclusion as Le Verrier, and at an earlier stage, they were furious; it was felt that Adams was trying to claim credit under false pretences. The affair nearly led to an international incident, but fortunately neither Adams nor Le Verrier took any part. When they met face to face, some time afterwards, they struck up an immediate friendship — even though Adams could not speak French and Le Verrier could not speak English.

Neptune proved to be very slightly smaller but appreciably more massive than Uranus. Its disk is bluish rather than green, and it does not share Uranus' peculiar axial tilt; its 'day' amounts to just over sixteen hours. In make-up it is very similar to Uranus, but differs inasmuch as it has a strong internal heat-source. Earth-based pictures could show clouds, and it was generally thought that Neptune would turn out to be a more active world than Uranus — as indeed it did, as was shown by the Voyager 2 fly-by of 1989. There are belts and spots, one of which, the Great Dark Spot, is as large relative to Neptune as the Great Red Spot is relative to Jupiter. Wind speeds on Neptune are very high.

Before the Voyager pass, two satellites were known. One, Triton, is large, and moves in a wrong-way or retrograde orbit; the other, Nereid, is much smaller, and has an orbit which is more like that of a comet than a satellite. Voyager discovered six new moons, all close in to the planet, together with dark, very faint rings. It was also confirmed that Neptune is a radio source, and has a magnetic field which is quite strong, though less intense than those of the other giant planets. Neptune is much too faint to be seen with the naked eye from Earth, though binoculars will show it. It takes over 164 years to complete one journey round the Sun, so that we have known it for less than one Neptunian 'year'!

limb, and there are winds, though naturally they have very little force.

Triton is multi-coloured. Its southern polar region, now in the middle of its long summer, is covered with what looks like pink snow, but the snow is not like ours; it is made up of solid nitrogen. The northern hemisphere is different, and darker in hue, with innumerable shallow enclosures, while there is a bluish band near the equator. There is almost no surface relief on Triton, and, as with Europa in Jupiter's system, the difference between the highest elevation and the deepest valley can be no more than a hundred feet at most.

Amazingly, there is activity. Below the frozen surface there is presumably a sea of liquid nitrogen, and when any of this liquid percolates upward, so that the pressure is relaxed, it explodes in a shower of nitrogen ice and vapour, producing what can only be called a geyser. The dark streaks in the pink southern snow tell us where these materials have been blown downwind in the thin atmosphere.

Who would have expected nitrogen geysers on Triton? Few astronomers would have believed it, and despite the evidence which we on Voyager 2 have been able to send back there are still some sceptics. At any rate, Triton is unique. Probably it is not a genuine satellite of Neptune at all, but was once an independent body which was captured in the remote past. This would also explain why it moves round Neptune in a wrong-way or retrograde direction.

What we would really like to do is to study Triton over a long period of time, and see what is happening there, but we have to admit that there is no chance of this. Within a few hours of the encounter we will be so far away that all details will be lost; remember, we passed over Neptune's pole at a relative speed of 60,000mph. All we can do is to take one last picture, showing the crescents of Triton and Neptune together.

You may ask why we cannot complete our tour by sending back close-range views of Pluto, the strange little world discovered as recently as 1930. The reason is that Pluto is in the wrong place. It has an orbit which is much less circular than those of the other planets, and between 1979 and 1999 it is closer in than Neptune: to survey it we would have to move inward toward the Sun once more, and we do not have nearly enough power. Neither can we use Neptune's gravity, because the mathematicians have worked out that we would have to burrow deep into Neptune's globe, which is out of the question. So Pluto must remain unexplored; we have no idea what it would be like.

Teatime delay

According to a report which seems to be reliable, James Challis was searching for Neptune when he saw an object which seemed to show a small disk. For some unknown reason he did not check at once, but meant to do so the following night. Unfortunately he stopped for a cup of tea on the way to the observatory, and by the time he arrived the sky had clouded over. Too late! Before he could observe again, Neptune had already been found by Galle and D'Arrest in Berlin.

1990 — A last look back at the Sun and its family, as we begin our never-ending journey into space. Our task is not done, and we will continue transmitting information as long as possible, but our main mission has been successfully completed. From our spacecraft we see the Sun as a dazzling point, but the planets have been almost lost to view, and the blackness around us is becoming more intense. Our cameras have been switched off, so that all our remaining power can be saved for the instruments which still function.

We can speculate as to what lies ahead. By 2020 we will reach the

Le Verrier and Paris

Le Verrier was not popular. In 1870 he was dismissed as director of the Paris Observatory because of his 'irritability', but he was reinstated when his successor, Delaunay, was drowned in a boating accident off the French coast.

Novice astronomer

When Clyde Tombaugh discovered Pluto he was a young student, who had been engaged by the Lowell Observatory specially for the search. Later he arrived at Kansas University to take an official degree. It is hardly surprising that the Director refused to let him enrol for the course in elementary astronomy!

boundary of the 'heliosphere', where the solar wind becomes undetectable. Around this time all contact with Earth will be lost, and our final fate will never be known. In time we will pass through the Oort cloud of comets which moves round the Sun at a distance of about a light-year, and by around AD295,000 we should be within reasonable range of Sirius, which will then appear brighter to us than Venus does from Earth.

Of course, there is always the danger of a collision with a wandering body, but there is also the chance that we will be found and collected by an alien civilisation. For this reason we carry a plaque, and also a record on which there are 'sounds of Earth'. Meanwhile, we on Voyager 2 will say goodbye, conscious that we have done our best.

I know that this has been nothing more than a journey in imagination, but if Voyager 2 had indeed carried any passengers they would have had much the view I have described. Bear in mind, too, that if all goes well we ought to be able to send astronauts to the outer Solar System within the next few centuries. Let us, then, pay tribute to Voyager 2. It is sad to see it leave us, and perhaps it will one day be on show in some alien museum!

PLUTO AND PLANET 'X'

Even when Neptune had been discovered, there was still something 'not quite right' about the movements of the outer planets. Percival Lowell, of Martian canal fame, came to the conclusion that there must be yet another planet, at a still greater distance from the Sun, and he began to make calculations with the hope of locating it. He failed to find it, but in 1930, fourteen years after Lowell's death, Clyde Tombaugh, working at the observatory which Lowell had founded at Flagstaff in Arizona, found the new planet not very far from the predicted position. Tombaugh — then a young amateur, now one of America's most senior and respected astronomers — was using a 13in refracting telescope which had been bought specially for the purpose.

Pluto, as the planet was named, has set astronomers problem after problem. Its orbit is much less circular than those of the other planets, and when at its nearest to the Sun it is closer in than Neptune, though its path is tilted at the relatively sharp angle of 17° and there is no fear of a collision. The last perihelion fell in 1989, so that between 1979 and 1999 Neptune, not Pluto, ranks as 'the outermost planet'. More significantly, Pluto is not a giant, and has turned out to be even smaller and less massive than the Moon. This means that it could not possibly pull a giant such as Uranus or Neptune measurably out of position. Unless Lowell's reasonably correct prediction was sheer luck, which is hard to believe, the real planet for which Tombaugh was searching is further away and has yet to be discovered. We usually refer to it as Planet X.

Pluto is not alone. It moves together with a companion, Charon, whose diameter is about half that of Pluto—753 miles, as against 1,444 miles for Pluto. Pluto spins on its axis in 6 days 9 hours, and the revolution period of Charon is exactly the same.

Pluto has a very thin methane atmosphere, with a density no more than one-millionth that of our air. It is now moving outward toward its next aphelion in 2116, and as the temperature falls it seems likely that the atmosphere will freeze on to the surface. Charon seems to have no atmosphere at all.

Pluto does not seem to 'fit in' with the general plan of the Solar System. There have even been suggestions that is used to be a satellite of Neptune which broke away for some reason, moving off in an independent orbit. This would at least explain why Triton now moves round Neptune in a wrong-way direction, because the departure of Pluto would have involved a tremendous disturbance of some kind, but we cannot be sure, and mathematicians have grave doubts about the whole idea. In any case, it seems that Pluto is unworthy of true planetary status. It and Charon might be better regarded as a double asteroid.

If Planet X exists, it is bound to be very faint and hard to find. Certainly it will be lonely and desolate beyond our understanding.

Tales of the Unexpected

Life is full of surprises. Some of them are pleasant, others not. Astronomy is no exception, and there is always the chance of being caught unawares by some startling event.

Auroræ, or polar lights, come into this category to some extent, because we can never predict them with any accuracy, and they can be truly magnificent. We know that they are caused by electrified particles sent out by the Sun, and they have been observed for many centuries; in Roman times, the Emperor Tiberius once dispatched his fire engines to the port of Ostia to put out what seemed like a dangerous blaze, but which proved to be nothing more alarming then a brilliant red aurora. We also know when auroræ are quite likely to occur, but we can never tell just where or when.

The Sun may be regarded as a variable star. Every eleven years it

The great aurora of 13 March 1989, showing a red ray crossing the constellation of Boötes (Paul Doherty)

Auroral noise

There have been many report of hissing and crackling sounds heard during auroral display. Since aurorae are high-altitude phenomena over 100 miles above the ground, these sounds are by no means easy to explain, and at best we must reserve judgement. Strange odours have also been reported, but smelly auroræ seem even less likely than noisy ones.

becomes active, with many sunspot groups and frequent flares. It is the flares which send out the electrified particles, which cross the 93,000,000 mile gap between the Sun and the Earth and enter the Van Allen radiation zones which surround our world. The Van Allen zones become so overloaded that particles cascade downward into the upper air to produce the lovely glows which are so common in latitudes such as those of Norway or the Shetland Isles (or, in the far south, Antarctica). Auroræ are commonest in high latitudes, because the electrified particles spiral toward the magnetic poles, but they can spread closer to the equator, and there is even one record of an aurora seen from Singapore.

Brilliant auroræ are uncommon from South England, but they do occur now and then. The best that I have ever seen was on 26 January 1938, when from my Sussex home the entire sky was bright orange in colour. Much more recently, on 13 March 1989, a spectacular display caused widespread interest over the whole country; from Stoke-on-Trent the astronomer-artist Paul Doherty telephoned me to

THE SOLAR CYCLE

In 1826 Heinrich Schwabe, a Dessau pharmacist with a keen love of astronomy, began to make drawings of sunspots on every clear day, projecting the image with a small telescope. He was not specially interested in the Sun, but he believed in the existence of a planet closer in than the orbit of Mercury (later christened Vulcan), and he hoped to catch sight of it as it passed in transit across the Sun's face.

He never saw Vulcan, which is not surprising in view of the fact that it does not exist, but he found something of much greater significance. To his surprise, the numbers of sunspots waxed and waned over a definite period of about eleven years. For example, there was great activity around 1837, 1848 and 1860 but at minimum, around 1843, 1856 and 1867, there were many consecutive days when the disk was completely blank. When Schwabe announced his findings, astronomers were somewhat sceptical, but they soon found that he was right, and it did not take long to establish connections between the solar cycle, displays of auroræ and 'magnetic storms' — that is to say sudden, rapid variations of the compass needle.

Later investigations brought out some even more curious points. Tree-rings, formed during a tree's slow growth, were found to be affected by the state of the solar cycle, and this new science of 'dendrochronology', pioneered by an astronomer named Andrew Douglass (one of Percival Lowell's colleagues) has proved to be of great value to climatologists.

There also seemed to be a period, between 1645 and 1715, when there were almost no spots at all, so that to all intents and purposes the solar cycle was suspended. Unfortunately the records are not complete, but the evidence is strong, and there was also a lack of reported auroræ, while during total solar eclipses the corona was much fainter than usual. This period is called the Maunder Minimum, because attention was drawn to it by the British astronomer E.W. Maunder. It was also marked by a spell so cold, in Europe at least, that it has been referred to as 'the Little Ice Age'; during the 1680s the Thames froze almost every year, and 'frost fairs' were held on it. At one stage the temperatures in Northern Europe were so low that there was serious talk of evacuating Iceland.

There may have been an earlier spotless period between 1400 and 1510, though the records are too fragmentary for us to be sure, but certainly the solar cycle is well marked at the present time; the last maximum was that of 1990, with minimum in 1986-7. The next minimum is due about 1995.

Whether the state of the cycle has any effect upon our day-to-day weather is problematical, but it is worth noting that the very hot British summers of 1975 and 1976 were at the time of solar minimum, while the equally blazing summers of 1989 and 1990 were at a time when the Sun was rising toward maximum. Neither do we know the real cause of the cycle, and we have to admit that we are very uncertain about the origin of sunspots themselves. Though the Sun is the only star which we can examine in detail, there is much that we do not know, and it may well be that we actually know less than we believed we did a few years ago!

ECLIPSES OF THE SUN

The fact that the Sun and the Moon appear so nearly the same size in our skies is pure coincidence — but it is lucky for us, as otherwise we would never know the glory of a total eclipse, and before the space Age we should have been ignorant of the existence of the corona.

Because the Moon's shadow is only just long enough to reach the Earth, one has to be in just the right place at just the right time to see a solar eclipse. This is not true of an eclipse of the Moon, which is visible from any part of the world over which the Moon happens to be above the horizon. Therefore, any particular location will see more lunar than solar eclipses, even though they are no more common when considered together. There must be at least two eclipses per year (both solar) and there may be as many as seven (five solar, two lunar).

Solar eclipses are of three kinds:

1. Total, when the sun's bright surface or photosphere is completely hidden. The track of totality is never more than 169 miles wide, and the maximum possible length of totality is 7½ minutes, though most eclipses are much briefer than this.
2 Partial, when only a portion of the sun is hidden, A partial eclipse is seen to either side of the zone of totality, but many partial eclipses are not total anywhere on the Earth.
3 Annular. When the alignment is perfect, but when the Moon is near apogee (its greatest distance from the Earth) it does not appear quite large enough to cover the Sun, and a ring of sunlight is left showing round the dark body of the Sun (Latin *annulus*, a ring). Annularity can last up to 12½ minutes.

continued overleaf

ECLIPSES OF THE SUN BETWEEN 1990 AND 2000 ARE AS FOLLOWS:

Date	Area	Type
1990 26 Jan	Antarctic	Annular
1990 22 July	Finland, Russia, Pacific	Total:2min33sec
1991 15-16 Jan	Australia, N Zealand, Pacific	Annular
1991 11 July	Pacific, Mexico, Brazil	Total: 6min 54sec
1992 4-5 Jan	Central Pacific	Annular
1992 30 June	South Atlantic	Total: 5 Min 20sec
1992 24 Dec	Arctic	Partial:84 per cent
1993 21 May	Arctic	Partial 74 per cent
1993 13 Nov	Antarctic	Partial: 93 per cent
1994 10 May	Pacific, Mexico, USA, Canada	Annular
1994 3 Nov	Peru, Brazil, S Atlantic	Total: 4min 23sec
1995 29 Apr	S Pacific, Peru, S Atlantic	Annular
1995 24 Oct	Iran, India, E Indies, Pacific	Total: 2min 5sec
1996 17 Apr	Antarctic	Partial: 88 per cent
1996 12 Oct	Arctic	Partial: 76 per cent
1997 9 Mar	Arctic, Russia	Total: 2min 50sec
1997 2Sept	Antarctic	Partial: 90 per cent
1998 26 Feb	Pacific, Atlantic	Total: 3min 56sec
1998 22Aug	Indian Ocean, E Indies, Pacific	Annular
1999 16 Feb	Indian Ocean, Australia, Pacific	Annular
1999 11 Aug	Atlantic, England, France, Turkey, India	Total: 2min 23 sec

In 1999 the track will cross Cornwall. Book your hotel early!

say that the aurora was casting strong shadows.

The last solar maximum fell as recently as 1990, so that we may well be treated to another major display. Of course, if you take the trouble to go up to, say, North Scotland, you will see an aurora of some kind on almost every clear night.

One phenomenon which took astronomers very much by surprise, and which caused them a great deal of annoyance, was that known as the Black Drop — and this brings me on to say something about transits of Venus.

Mercury and Venus, remember, are the only two planets closer to the Sun than we are. When they pass directly between the Sun and the Earth, they appear as black disks against the solar face. Mercury does so reasonably often (the last transit was in 1993, the next will be in 1999), but Venus is much less obliging. Transits take place in pairs; there is one pair, separated by eight years, and then no more for over a century. Thus there were transits in 1874 and 1882, but the next will not be until 2004 and 2012.

There can be nobody now alive who can remember a transit of Venus, but we have many descriptions of them, and photographs as well. A complete transit takes some hours, and the planet is clearly visible with the naked eye as a black spot, much darker than a sunspot group. (A sunspot is not really dark at all; it only appears so because its temperature is some 2,000°C below that of the surrounding surface. If it could be seen shining on its own, its brilliance

continued

Total eclipses are much the most interesting, because only then can the Sun's surroundings — the chromosphere, prominences and corona — be seen with the naked eye. The chromosphere ('coloursphere') is the lower atmosphere, and into it rise the masses of red gas which we call prominences. They are made up chiefly of hydrogen, and are magnificent spectacles. The slightest segment of the Sun's bright surface will drown them, though with special equipment they can now be observed at any time.

The main feature of totality is the pearly corona, which spreads out in all directions, sometimes more or less symmetrically and sometimes with long rays (it depends mainly upon the state of the solar cycle). The corona gives out about as much light as the full moon, but there is one very important point to be borne in mind. During actual totality it is quite safe to look at the Sun, or to use the lens of an SIR camera for photography, but even a tiny portion of the bright surface is highly dangerous — so be careful not to have your eye to a camera lens, binoculars or a telescope when totality is about to end. (Also, beware of looking straight at a partially eclipsed Sun, which is to all intents and purposes as bad as looking straight at the Sun in normal times. At every eclipse there are cases of people who have had to go to hospital for urgent treatment because they have neglected this warning.)

Nobody who has seen a total solar eclipse is likely to forget it. Just before the onset of totality there is a sudden drop in temperature; the Moon's shadow comes racing toward you at a fantastic speed, and you may see 'shadow bands', wavy lines across the Earth's surface, which are purely atmospheric phenomena. (Try to photograph them if you have time; it is unexpectedly difficult.) Then, as the last sliver of the Sun disappears, there is the wonderful 'diamond ring' effect, together with bright spots round the Moon's edge known as Baily's Beads; they are caused by the sunlight shining through lunar valleys. Then, suddenly, the sky darkens, and the chromosphere, corona and prominences flash into view. It seems as though Nature comes to a halt; the wind drops, and there is an abrupt silence. It has even been said that flowers close up and birds roost (thought I admit that I have never been too sure about the birds). The moments of totality race by; then there is the return of the diamond ring, and as the Sun reappears the brilliant corona vanishes.

An eclipse expedition is immensely enjoyable. If you ever have the chance to join one, I strongly advise you not to pass it by.

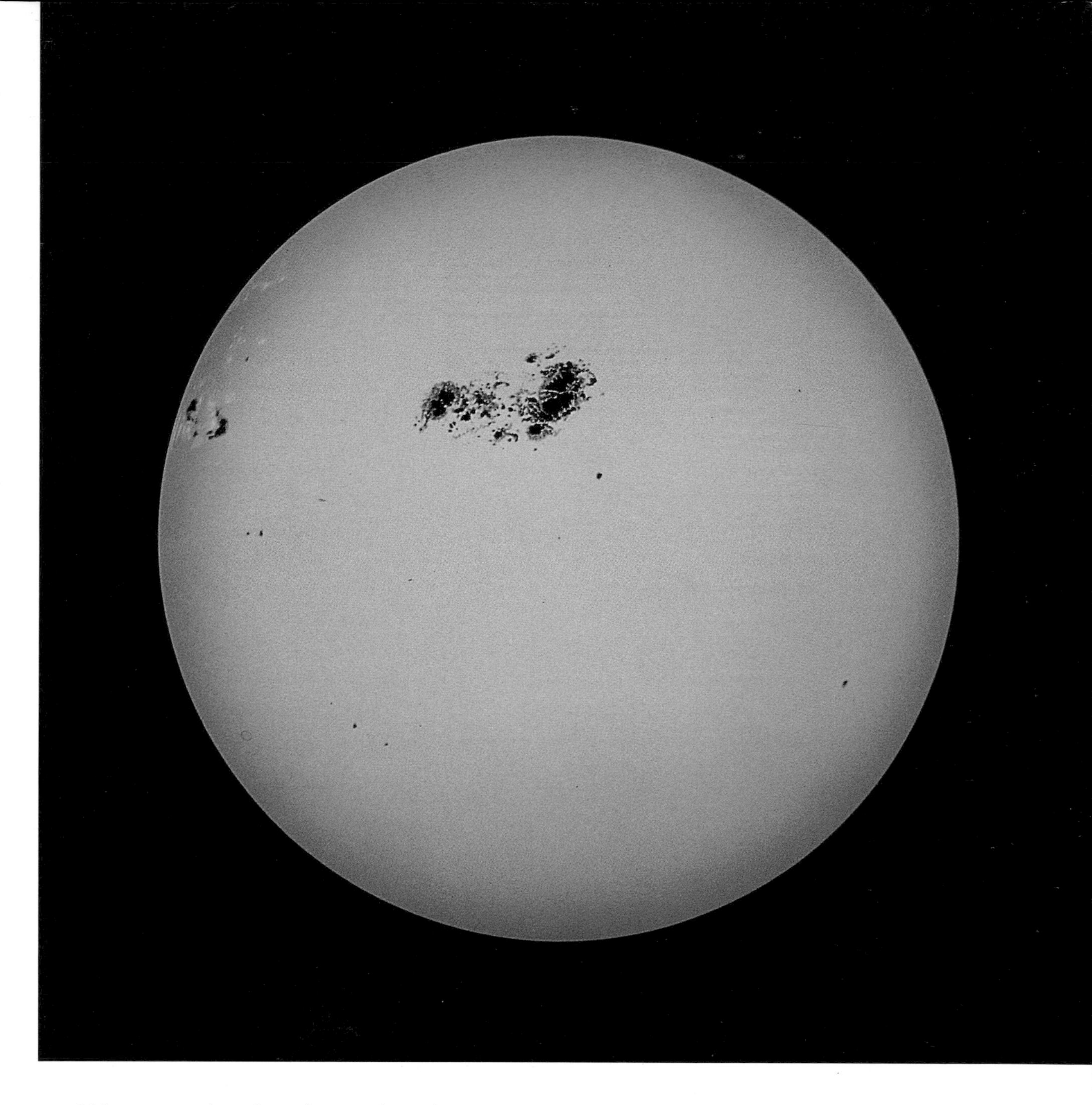

would be greater than that of an arc-lamp.)

Transits of Venus used to be regarded as very important, because they provided a method of measuring the distance between the Earth and the Sun. What had to be done was to time the exact moment when the transit began, and also the moment when it ended. In 1874 expeditions were dispatched to various parts of the world, but the results were poor, because when Venus moved on to the Sun it seemed to draw a strip of blackness after it; when this strip disappeared, the transit was already well under way. The so-called Black Drop ruined the accuracy of the whole method, though there are now much better ways of measuring the distance of the Sun, and the transits of 2004 and 2012 will be regarded as no more than casually interesting.

Great sunspot of April 1947, the largest ever observed. Easily visible with the naked eye, this drawing by Paul Doherty was made from my own observatory of it

There is no particular mystery about the cause of the Black Drop. It is due to Venus' dense, extensive atmosphere. But the phenomenon was certainly unexpected, and I look forward to seeing it for myself in 2004.

Dust storms on Mars can cause trouble, too; when the probe Mariner 9 neared its target, in 1971, all that could be made out was the almost featureless top of a dusty layer. The only features which

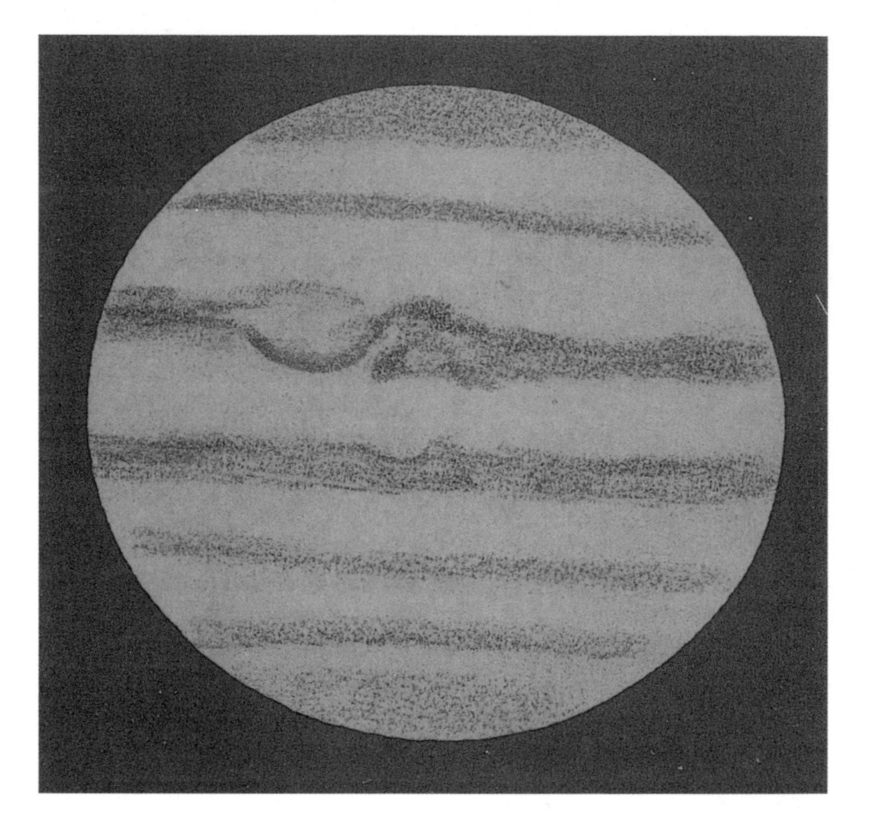

Jupiter, 10 October 1988. 15in reflector x 400. South is at the top, the South Equatorial Belt is the most prominent of all the figures on the disk

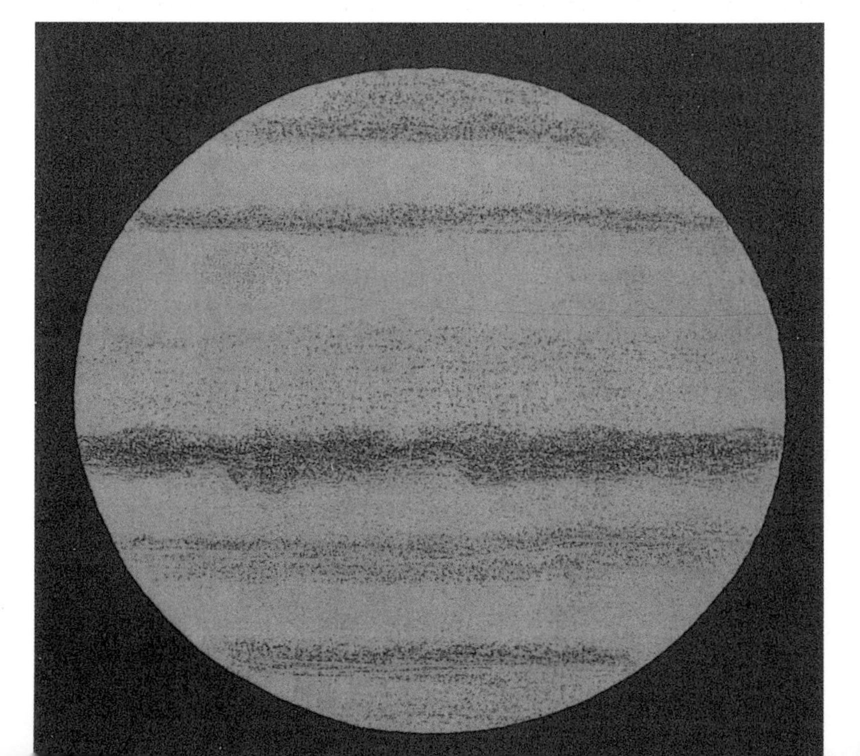

Jupiter, 29 November 1989. Now the South Equatorial Belt has virtually disappeared, but the Great Red Spot is returning. 15in reflector x 400

The white spot on Saturn, discovered in 1933 by W. T. Hay. This is his own drawing of it

could be seen were patches marking the tops of giant volcanoes poking out above the general murk, and it took some time for the dust to clear away to allow Mariner to begin its survey.

Jupiter, the giant of the Sun's family, is always changing. The Red Spot can sometimes disappear for a while, though it always comes back — and in mid-1989 one of the main belts, the South Equatorial, suddenly vanished. When I first looked at Jupiter when it came out from behind the Sun in the early hours of the morning, I thought for a moment that there was something wrong either with my telescope or with my eyes. The belt reappeared in late 1990.

Saturn provided a considerable shock in 1933, when a bright white spot near its equator was discovered by a most unusual astronomer — W.T. Hay, better remembered today as Will Hay, the stage and screen comedian, who was a skilful and enthusiastic amateur observer. The spot did not last for long, but it marked a violent upheaval in the planet's upper clouds. I well remember seeing it with my newly acquired 3in telescope.Another white spot appeared in 1990.

Even the Moon can surprise us. Lunar observation has changed over the past thirty years, because of the spacecraft results, but until 1957 or thereabouts the main emphasis was on mapping the surface, and this was my own particular line of research. Just after the end of World War II, I was looking at one particular part of the lunar limb when I detected what I thought might be a small sea or 'mare'. It was not on my map; it was hard to see, because it was so foreshortened, and it came into view only when conditions were exactly right. Together with a colleague, Percy Wilkins (a Civil Servant by profession, an astronomer by inclination) I charted it, sent in a report to the British Astronomical Association, and even suggested a name for it: Mare Orientale, or the Eastern Sea, because it was on the eastern edge of the Moon. Later, the space-probes showed it to be a vast, ringed enclosure extending on to that part of the Moon which we can never see from Earth because it is always turned away from us. Today it is recognised as being one of the most significant features on the whole of the lunar surface.

East is west

I suggested the name Mare Orientale (the Eastern Sea) because the feature was on the Moon's eastern edge. In 1966, however, the International Astronomical Union held a meeting in Czechoslovakia, and decided to reverse lunar east and west. I opposed the change, but when it came to a vote I was heavily defeated, so that my *Eastern* Sea is now on the Moon's *western* limb.

It is usually said that the Moon is a dead world, where nothing ever happens. This is not quite true, because now and then we can see occasional signs of activity. In 1958 the Russian astronomer Nikolai

Mare Orientale from Orbiter 4. Now we see that the Mare is a vast ringed formation, extending well on to the Moon's far side

Kozyrev, using a large telescope at the Crimean Astrophysical Observatory, detected a red glow in the large walled plain Alphonsus, and attributed it to a volcanic eruption. Conventionally minded astronomers were taken aback, but amateur observers, who had been studying the Moon for years, were less sceptical. By now the reality of these elusive outbreaks is hardly in doubt, and it seems that they are due to the release of trapped gases from below the lunar crust. I have seen a number of them myself.

So far as the Solar System is concerned, the main surprises come from comets and meteoroids. Brilliant meteors can be startling, and they can never be predicted. Occasionally a meteorite is seen to fall; this was the case on Christmas Eve, 1965, when an object shot across the skies of England and a meteorite crashed to earth near the Leicestershire village of Barwell. It broke up before landing, and showered fragments over a wide area, many of which were later collected. One piece came through the open window of a house, and was later found nestling coyly in a vase of artificial flowers.

Comets are the vagabonds of the Sun's family, and are always erratic. This applies to some extent even to the most famous of all comets, Halley's, which was last on view with the naked eye in 1986, and will be with us once more in the year 2061.

The name honours Edmond Halley, who was England's second Astronomer Royal. He was a close friend of Isaac Newton, and indeed it was he who persuaded Newton to write his great book about gravitation. He made many notable contributions to astronomy, but he is best remembered today because of his association with comets, which were at that time regarded as decidedly mysterious; even Newton thought that they moved in straight lines, so that no comet could pass by the Sun more than once.

In 1682 a bright comet appeared, and Halley observed it carefully. When he worked out its path, he found that it was moving in much the same orbit as comets previously seen in 1531 and in 1607. Could the three bodies be one and the same? Halley believed so, and wrote: 'You see, therefore, an agreement of all the elements in these three, which would be next to a miracle if they were three different comets; or if it was not the approach of the same comet towards the Sun and Earth in three different revolutions, in an ellipsis around them. Wherefore, if accordingly to what we have already said, it should return again in the year 1758, candid posterity will not refuse to acknowledge that this was first discovered by an Englishman.'

The comet did return, though Halley did not live to see it (he died in 1742). Subsequently it has been traced back through history, and has appeared every seventy-six years or thereabouts. For example, it shone down in the year 1066, not long before the Battle of Hastings, and caused considerable alarm among the Saxons, who regarded it as an evil omen. Chinese records of it go back well before the time of Christ.

It was seen again in 1835, and was observed throughout the world; actually, the last sighting of it was made by Sir John Herschel during his sojourn at the Cape. The next return, that of 1910, was eagerly

Opposite:
Comet of 1882. This was the famous picture taken by David Gill at the Cape, showing so many stars that he realised that the best way to map the sky was by photography

awaited. The comet did not disappoint us, but it was completely upstaged. In January, several weeks before Halley became bright enough to be seen with the naked eye, a group of diamond miners in South Africa looked upward when they emerged from their spell of duty, and were astounded to see a bright comet shining down at them. Later in the month it became so brilliant that it remained visible even when the Sun was above the horizon, so that it was nicknamed the Daylight Comet. As it faded, Halley brightened up to reach its maximum in the following April; but though it became quite prominent, it never rivalled its predecessor.

The Daylight Comet moves in a very long, narrow path. Its estimated period of revolution is of the order of 4,000,000 years, and though this may be highly inaccurate we may be sure that it will not return for a very long time. Halley, however, came back on schedule in 1986, and provided astronomers with another major surprise.

Halley's Comet can be really spectacular. At the return of AD837

HALLEY AND NEWTON

Edmond Halley is one of the most famous astronomers in history. Though he is best remembered as being the first man to predict the return of a comet, this was only one of his many achievements — and he was a colourful character as well.

He was born in London in 1656, son of a successful businessman. There was no shortage of money, and he went to St Paul's, where he became school captain, before entering Queen's College, Oxford, to study for his degree. Before graduating, however, he went to the island of St Helena to make the first really systematic survey of the southern stars, which never rise over Europe and were poorly known. He had problems, both with the weather and with a particularly disagreeable island governor named Gregory Field, but by the time he came home he had done all that he had hoped to do, and had produced an excellent catalogue of 300 stars. His reputation was made, and he was awarded an honorary degree by Oxford University, as well as being elected a Fellow of the prestigious Royal Society.

By now he was in touch with most of the leading scientists of the day, including Isaac Newton, Sir Christopher Wren (who was a professional astronomer before turning to architecture) and the brilliant but cantankerous Robert Hooke. In 1684 Halley, Wren and Hooke had long discussions about the nature of gravitation, and in particular what we now call the 'inverse square law'. This law seemed to be valid, but even Halley and Hooke, who were excellent mathematicians, could give no proof. Only one man was capable of doing that: Isaac Newton.

Newton was in Cambridge, and was not a 'companionable' man; he was what in modern parlance might be termed a loner. When Halley journeyed to Cambridge to consult him, he found, to his amazement, that Newton had solved the problem years earlier, but had never published his results, and had even lost his notes! Halley persuaded him to re-work the calculations, and then to write the immortal book which we now call the *Principia*. Newton was far from enthusiastic, but even when the work was finished there were difficulties about publishing it; the Royal Society had no money to spare. So it was published at Halley's personal expense, and it caused a complete revolution in scientific outlook. It has been described as 'the greatest mental effort ever made by one man', but without Halley it would never have been written.

In 1720 Halley succeeded John Flamsteed as Astronomer Royal at Greenwich Observatory, and remained there for the rest of his life. Among his discoveries was that of stellar 'proper motion'; he found that several bright stars, including Sirius, had moved perceptibly against their background since ancient times. Quite apart from his astronomical work, he carried out long sea voyages to study the lines of magnetic force, he was called in as adviser with respect to military fortifications in Europe, and he even descended below sea-level, off the Sussex coast, in what can only be described as a primitive submarine. Clearly he was very much of an 'all-rounder'.

Halley was a kindly, jovial man with a strong sense of humour. He was healthy almost to the last, and typically, his last action, only minutes before he died, was to call for a glass of wine — and drink it.

Meteors from Biela's Comet, 1872. This is one of the only sketches made of the meteor shower which marked the funeral pyre of Biela's Comet

Halley and the Czar

Halley was quite unlike the austere Astronomer Royal, John Flamsteed, who commented acidly that Halley 'swore and drank brandy like a sea-captain', which he probably did. When the fearsome Czar of Russia, Peter the Great, came to England to learn about shipbuilding he struck up a friendship with Halley, and it is said that on one occasion, after a far from teetotal evening, the Czar climbed into a wheelbarrow and Halley pushed him through a hedge. This may or may not be true, but certainly the Greenwich records contain Halley's receipt for a damaged wheelbarrow. . .

it by-passed the Earth at a distance of less than 4,000,000 miles, and according to contemporary accounts it developed a tail over 100° long, with a head shining more brilliantly than Venus. Things were different in 1986, because, by sheer ill-fortune, the comet was badly placed in the sky; when it ought to have been at its best, it was on the opposite side of the Sun and could not be seen at all. It did become a naked-eye object, but many people were disappointed by it — and I have lost count of the number of fairly elderly people who told me, proudly, that they remembered Halley in 1910; I never had the heart to tell them that in all probability they had seen the Daylight Comet instead!

Yet this latest return was of tremendous importance, because for the first time it was possible to explore Halley's Comet by using space-craft. Preparations were made well ahead of time. The Americans cancelled their Halley mission because it would cost too much (though, please note, they could have built several comet probes simply by cancelling one nuclear submarine), and the encounters were left to the Japanese and the Russians, with two space-craft each, and the Europeans, with one. All five were scheduled to rendezvous

Where are they now?

When John Flamsteed became Astronomer Royal he had to provide his own instruments, and when he died Mrs Flamsteed descended upon Greenwich Observatory like an east wind, removing all the equipment — which was legally hers. When Halley became Astronomer Royal, therefore, he had to begin all over again. The sad part of this story is that the original Flamsteed instruments cannot now be traced.

The story of Stellaland

There was once an independent nation named after a comet! In 1882 two small republics appeared in Southern Africa, in the area now occupied by Botswana. One, Goschen, had its capital at Mafeking. The other was centred round Vryberg, and a name had to be found for it; why not call it after the bright comet then visible in the sky? So the name chosen was 'Stellaland'.

Nobody seems to have much idea who controlled and organised it, but clearly it did no harm to anyone, and it issued stamps, countersigned by one J.P.Minaar, the Treasurer-General. After three years there came a crisis. Cecil Rhodes was organising the Cape to Cairo Railway, and Goschen and Stellaland were in the way. This could not be allowed; the Cape authorities dispatched a force under Sir Charles Warren to deal with the situation, and when it arrived Stellaland ceased to exist. Not a shot was fired, but the little republic was officially ended on 7 February 1885.

with the comet in late March 1986, and one of them, Europe's 'Giotto', was planned to go right into the comet's head and take close-range pictures of the nucleus.

Up to that time, nobody had ever had a proper view of the nucleus of a comet. At a great distance from the Sun there is nothing much to be seen, while as the comet moves inward the ices evaporate to such an extent that the nucleus is hidden behind an opaque veil. So what would 'the heart of Halley' be like? Most astronomers expected it to be bright and icy, but a few, notably Sir Fred Hoyle and his Indian colleague Chandra Wickramasinghe, did not; they believed that the nucleus would be dark, with the icy core covered up by blackish, possibly organic material.

To the consternation of most theorists, Hoyle and Wickramasinghe proved to be absolutely right. The Halley nucleus was black, and there were only isolated 'fountains' through which the icy material was spurting out from below. It was a real shock; I wonder what Edmond Halley would have thought about it?

As the comet moved away it faded; it was still trackable in 1990 with the world's largest telescopes, but before long we will lose it, not to see it again until it draws in toward its next perihelion passage in 2061. Undoubtedly it will be on schedule — but there have been cases of other comets which have astounded astronomers not by appearing 'out of the blue', but by failing to do so.

In 1826 an Austrian army officer, Wilhelm von Biela, who was also a keen amateur astronomer, discovered a comet. Calculations showed that it moved round the Sun in a path much smaller than that of Halley's Comet, and would be expected to return every 6¾ years. It duly appeared in 1832, and even caused something of a panic in France, because there had been a report that it would collide with the Earth. (Actually, it never came within ten million miles of us, and in any case even major comets are so flimsy that even a direct hit would not destroy the world, although certainly it would do a great deal of damage.) Biela's Comet was on the fringe of naked-eye visibility in 1832, but it was missed at the next return, that of 1839, because it was badly placed in the sky. It was back again in 1845, and then came its remarkable metamorphosis. Without the slightest warning, the comet split into two parts.

What could this mean? Astronomers were not sure, but there could be no doubt that the comet had started to break up, and gradually the two fragments drifted further apart. The next return, that of 1852, was eagerly awaited. Back they came, still together and still travelling round the Sun in tandem. It was all very peculiar.

Nothing was expected in 1859 because, as in 1839, the comet(s) lay in the wrong part of the sky; but in 1866 conditions should have been ideal, and searches began well ahead of time. They went on — and on — and on. No comet appeared, and in fact Biela's Comet has never been seen from that day to this. It has completely disappeared.

Yet it did not just 'softly and silently vanish away', like the hunter of the Snark. In 1872, when the comet was next due, a brilliant shower of meteors was seen coming from that part of the sky where

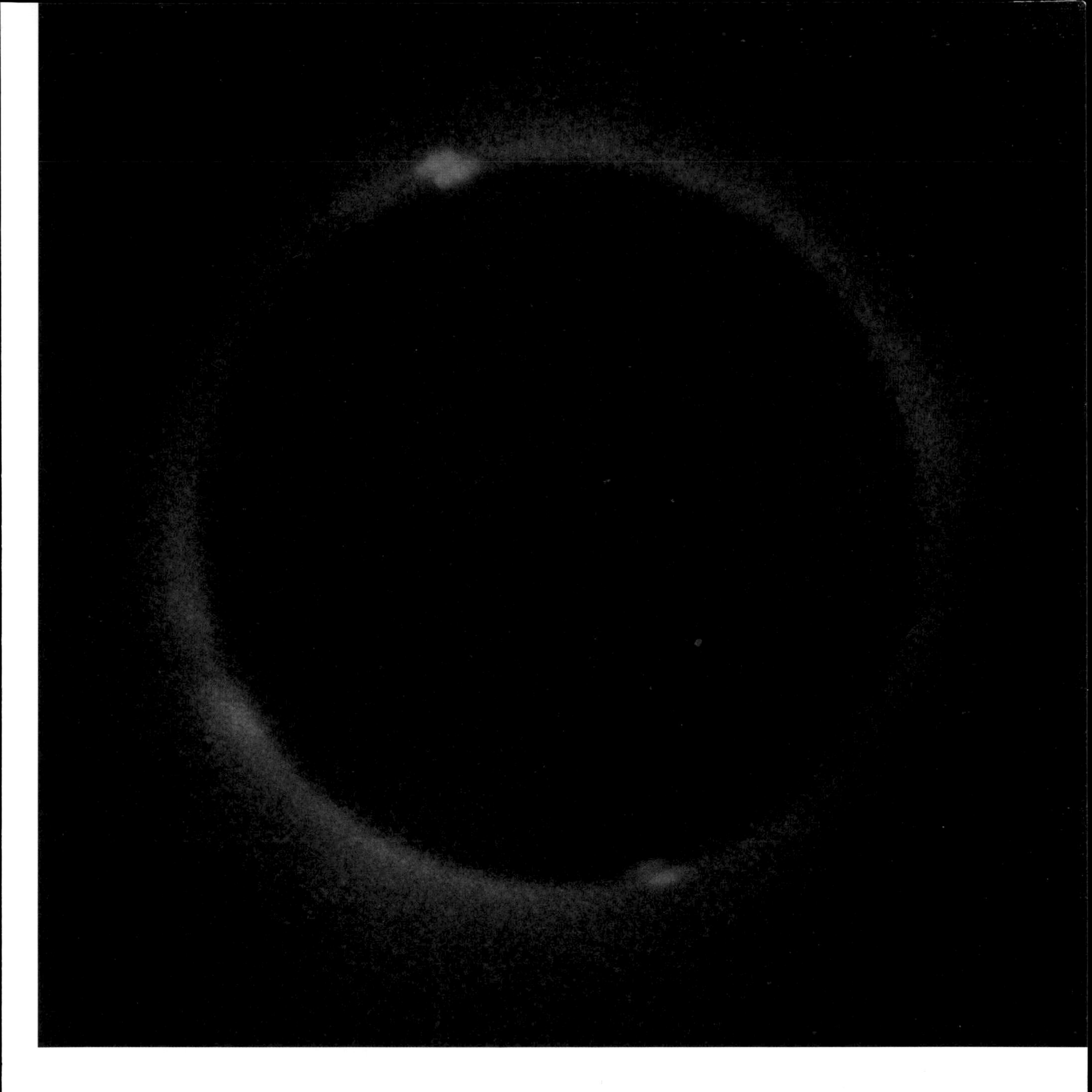

Total eclipse of the Sun, 17 March 1988. Taken from Talikud Island in the Philippines, under difficult conditions (rain was falling five minutes before totality). A couple of prominences are recorded

Biela's Comet should have been. There can be little doubt that they marked the funeral pyre of the comet, and there was another similar display in 1884, again when the comet ought to have been on view. Since then the supply of 'Bieliid' meteors seems to have become exhausted, so that by now we have probably seen the last of the dead comet in any form.

In 1882, when another bright comet appeared, Sir David Gill, Director of the Cape Observatory in South Africa, decided to try some photography. Photographic processes were very primitive in those days, and nobody had ever managed to obtain a satisfactory record of an object as ghostly as a comet, but Gill was nothing if not determined. He enlisted the aid of a photographer friend, and they took a time exposure by fixing a camera on to the main Cape telescope. When the plate was developed and printed, Gill saw to his astonishment that as well as giving a good view of the comet, the picture showed hundreds and hundreds of stars. At once he realised that the best way to map the sky was by photography, not by the laborious visual methods which had been used up to that time. So the photograph of the 1882 comet led directly on to the vast, detailed photographic sky atlases of today.

Yet another comet of 1882 was equally surprising, but for different reasons. It was observed only once — during a total eclipse of the Sun.

The Earth moves round the Sun; the Moon moves round the Earth.

PERIODICAL COMETS

Most astronomers, though not all, believe that comets are bona-fide members of the Solar System. Their masses are very slight; the only substantial part of even a bright comet is its nucleus, which is no more than a few miles across, and is icy in nature. When a comet passes close to a planet it is violently 'perturbed', and may have its orbit completely changed.

Comets seem to be very ancient objects, dating back to the early stages of the Solar System. According to the Dutch astronomer Jan Oort there is a whole 'cloud' of them, moving round the Sun at a distance of about a light-year (roughly six million million miles), much too far away to be seen from Earth. When a comet is perturbed for any reason, it may leave the Cloud and start to swing inward toward the Sun. After a journey lasting for many thousands of years it enters the inner part of the Solar System, and the ices in the nucleus begin to evaporate, producing a head or coma and (sometimes) a tail or tails.

One of several things may then happen. The comet may simply swing round the Sun and return to the Oort Cloud, not to be seen again for centuries. It may be 'captured' by a planet, usually Jupiter, and forced into a short-period orbit. It may be so affected by the planet that it is thrown out of the Solar System altogether. Or it may plunge to destruction in the Sun's atmosphere, as several comets have been observed to do.

Since a comet loses material every time it swings past the Sun, all the short-period comets, which have passed through perihelion many times, have exhausted most of their gases and have become faint; not many of them have tails, and very few ever reach naked-eye visibility (Halley's being the only exception, because it came in from the Oort Cloud in comparatively recent times by cosmical standards). The really spectacular comets, such as those of 1811 and 1843, encounter the Sun much less frequently, so that they have not lost nearly so much of their material.

Halley's was the first comet whose return was predicted. By now dozens of periodical comets are known, among them:

Name	First seen	Period (years)
Encke	1786	3.3
Grigg-Skjellerup	1902	5.1
Tempel 2	1873	5.3
Pons-Winnecke	1819	6.4
D'Arrest	1851	6.4
Kopff	1906	6.4
Schwassmann-Wachmann 2	1929	6.5
Giacobini-Zinner	1900	6.6
Borrelly	1905	6.8
Brooks 2	1899	6.9
Finlay	1886	7.0
Faye	1843	7.4
Tuttle	1790	13.7
Crommelin	1818	27.4
Brorsen-Metcalf	1847	72.0
Halley	Before Christ	76.0
Swift-Tuttle	1862	130 .0

Only Encke's has been seen at more than fifty returns (in fact, its small orbit allows it to be followed all the time, even when it is furthest from the Sun).

Quite a number of periodical comets have periods of around six to seven years. This is because their eccentric orbits take them out about as far as the path of Jupiter, and they are often referred to as Jupiter's 'comet family'. There was an old theory that they were actually shot out from Jupiter, though we now know that this is quite impossible.

Biela's is not the only periodical comet known to have 'died' in recent times, and it is true that these strange, ghostly objects must be short-lived. But great comets, with their long orbits, survive for much longer, and we may at any time see a new brilliant visitor coming in from the Oort Cloud.

This bright comet was first seen on these photographs of the total solar eclipse of 1947. Unlike Tewfik's Comet of 1882, it was also observed after the eclipse

Therefore, there are times when the three bodies move into a straight line, with the Moon in the mid position. By sheer chance, the Sun and Moon appear virtually the same size in the sky; the Sun is four hundred times further away, but its diameter is also four hundred times as great. During totality, then, the Moon can just blot out the brilliant solar disk, and for a brief period — never as long as eight minutes — the sky becomes dark; planets and bright stars shine out, and we see the Sun's surroundings, notably the masses of gas once (wrongly) called Red Flames and now known as prominences, and the glorious pearly corona, which is composed of rarefied gas and makes up what may be termed the Sun's outer atmosphere.

Before the Space Age, total solar eclipses were regarded as tremendously important, because they provided the only opportunities for making detailed studies of the corona. Unfortunately, the Moon's cone of shadow only just grazes the Earth, so that to see a total eclipse you have to be in exactly the right place at exactly the right time. On 17 May 1882 the central track crossed Egypt, and a party of astronomers made their way there. As soon as the last segment of the Sun's brilliant disk was covered, the watchers saw the corona — and also, to their astonishment, a comet! There was no previous record of it, and it was never seen again, so that we know absolutely nothing about its orbit. It is listed officially as Tewfik's Comet, in honour of the then ruler of Egypt.

A comet which passes very close to the Sun is in grave danger, and may well be destroyed. Such was the fate of a comet of 1978, Howard-Kooman-Michels, which was photographed from a spacecraft just before it plunged to destruction in the Sun's atmosphere. Other comets have been known to die in the same manner.

So far as great comets are concerned, the twentieth century has been rather barren; seldom have we had visitors to rival the really spectacular comets of the past, such as those of 1811, 1843, 1858 or 1882. But we never know when the next one will appear, and astronomers will certainly welcome it. There can be few sights to rival that of a brilliant comet, with a gleaming head and a long, shining tail stretching across the darkened sky.

Messier and Halley's Comet

In 1758, when the return of Halley's Comet was expected, Charles Messier was employed in Paris as Astronomer to the Navy, under the supervision of Nicholas de l'Isle. Naturally he was anxious to search for the comet, and he began well ahead of time, though he was hampered by persistent cloud. On 21 January 1759 he found the comet, and, of course, reported to de l'Isle who, for some unknown reason, ordered him to continue observing it, but not to announce the discovery! It was not until the following April that the announcement was made.

Messier would not have been the first in any case, because the comet was picked up on Christmas Night 1758 by the Dresden amateur Johann Palitzsch, but later the English astronomer J.R. Hind expressed the hope that 'such a discreditable and selfish concealment of an interesting discovery is not likely again to sully the annals of Astronomy'.

Astronomers in Strange Haunts

If you wanted to see the world's most modern and most sophisticated telescope, would you go to the Atacama Desert of Northern Chile? And if you felt inclined to make special studies of the Sun, would you go down into a gold-mine a mile beneath ground-level in the Black Hills of South Dakota, once the land of Calamity Jane, Wyatt Earp and Doc Holliday? The probable answer is 'no', but you would be wrong. This is just what astronomers are inclined to do.

'Astronomical travelling' is something which fits in with a passion for the stars. Take the La Silla Observatory in the Atacama Desert, for example. Think of a desert, and you will probably conjure up the picture of a sandy Sahara, with romantic Bedouins galloping about on camels amid the rolling dunes. The Atacama is not like that. There is no sand, but there are countless rocks and what can only be called 'dirt', with scrubby vegetation which clearly has a hard struggle to survive. Rainfall is rare. Snow is to be seen on the mountain-tops of the Andes, but at 8,000ft above sea-level, where La Silla has been built, snow is rare. For most of the time the daytime sky is blue and the night sky is velvety black, so that you feel as though you could almost reach up and pull the stars down.

The region is not exactly over-populated. The nearest town, La Serena, is over sixty miles away, and after that there is nothing much before you come to Santiago, Chile's capital. Around the Observatory there is a vast area of emptiness. On the drive from La Serena, up a steep, mainly metalled road, you find nothing but the occasional Indian village, if indeed you can call a sparse collection of huts a 'village'. Now and then you may catch sight of a large bird overhead; the condors are less common than they used to be, but you can still see them if you are lucky. On the ground, animal life is conspicuous only by its absence. Desert foxes survive (one of them often comes up to the Observatory near breakfast-time) and there are a few donkeys, plus insects such as scorpions which, fortunately, I have so far managed to avoid. But the overall impression of this part of the Atacama is one of desolation, and during the daytime it is very hot indeed.

Incidentally, there are frequent ground tremors. On my last visit to

Mauna Loa in action

From the summit of Mauna Kea, Mauna Loa is truly impressive; it is remarkable that of these two giant volcanoes, one is extinct while the other is so active. Mauna Loa can erupt with tremendous violence, and on one occasion its lava rolled as far as the outskirts of Hilo, threatening the town. A powerful local witch-doctor was called in and cast his very best spells; the lava halted before it could do any real harm, and, needless to say, legend has it that the witch-doctor was solely responsible.

Aerial view of La Silla, 1989 (Alastair Mitchell)

La Silla purchase

The land belonging to the La Silla Observatory covers a wide area: just over 240,000 square miles. It was purchased in the 1960s for the princely sum of $10,000. Even allowing for the fact that it is not the most fertile place on earth, this seems to be a fairly good bargain.

The NTT (New Technology Telescope) at La Silla, 1989

La Silla, in November 1989, I was woken at 3am by what was quite definitely an earthquake. The sound was rather like that of a heavy train passing by, and the windows and doors rattled for a few seconds. There was no damage, but earthquakes have to be borne in mind by astronomers who design telescopes.

There are two ways of reaching La Silla. One is by bus from Santiago, which is not to be recommended. The other is by air, either by a commercial flight to La Serena or, if you are an official visitor, by the European Southern Observatory's own small 'plane, which can come down on a small airfield within sight of the Observatory. The last part of the journey must be by road. From a distance, the Observatory looks like a collection of mushrooms sprouting up from the desert. The domes glint in the sunlight, and the contrast between the ultra-modern Observatory buildings and the ancient desert could not be more marked.

The NTT 'observatory'; the entire building rotates with the telescope

Night comes quickly at La Silla; once the Sun has set, which it does with surprising speed, darkness falls almost like a cloak, and from a suitable vantage point you can trace the Earth's shadow. La Silla at night is really pitch-black. Even La Serena is so far away that it can cause no irritating glow in the sky.

Of course, scattered light is one of the worst enemies of the modern astronomer, and a careless flash can do immense damage to an experiment which is designed to pick up the radiation from an object which may be thousands of millions of light-years away. Even car headlights are banned after dark, and considerable care is needed if you are walking along the road.(There are other hazards, too. Recently one astronomer, driving down the hill after an observing session, nearly ran into a wandering donkey.) La Silla is also silent, though now and then comes the grinding noise of a dome being rotated to give the observer the correct view of the sky.

I have mentioned the world's latest and most modern telescope – and here it is, on an eminence at the top of the main ridge of La Silla. It is known as the NTT or New Technology Telescope, and it was brought into full operation in 1989. It looks unfamiliar; for one thing it is housed not in a graceful dome, but in a strange, square building which rotates with the telescope itself, so that the slit in the building is always in the right position. The telescope is short and squat, and has no tube. The mirror, 3.58m (158in) across, is very thin, and this causes potential problems, because the mirror tends to distort under its own weight as the telescope is moved around to point at different objects in the sky. Therefore, the NTT is fitted with special 'correctors' behind the mirror which automatically put it back into its proper shape. This technique is called active optics, and improves things tremendously.

But why come to La Silla at all? The site itself is splendid, with no light pollution, but there is another equally important factor: height. Hillary and Tenzing climbed Everest because it was there. Astronomers go up mountains not because they like climbing, but because they are anxious to observe from above the thickest part of the Earth's atmosphere. Star-twinkling may look beautiful, but to the

The very large telescope

An even larger telescope is now being planned. In fact, the VLT will be made up of four 8-metre telescopes working together, providing a light grasp equal to that of a single 16m or 438in mirror.

This will be set up at Cerro Paranal, well to the north of La Silla but still in the Atacama Desert. The telescope should be working within the next ten years.

astronomer it is an unmitigated nuisance. What is needed is a steady image; and the less the amount of air above you, the less unsteadiness there is likely to be. Moreover, there is less water vapour to absorb the longer wavelength radiations which we term infra-red. An astronomer working in this field of research likes to be as 'dry' as possible.

So the answer is — Go aloft, young man! And so far as major observatories are concerned, the loftiest of all is to be found at an altitude of around 14,000ft, on the summit of Mauna Kea in Hawaii.

If you have never been so Hawaii, you may think that it is all yellow sand, surfing and bikinis. Part of it is, and Honolulu, on Oahu, is simply a large and noisy American city. But Big Island is different. There is only one town of any size, Hilo; the sand is black and volcanic, and there are two of the world's largest shield volcanoes, Mauna Loa and Mauna Kea. Mauna Loa is very active, and on its site is Halemaumau, which is worth seeing even when quiescent, while

The mighty keck

The Keck Telescope, on Mauna Kea, has a 396in mirror made up of 36 segments fitted together to make the correct curve. A twin is being planned, and will be erected beside it. When the two Kecks are working together, they will in theory be capable of detecting a car's headlights separately from a distance of 16,000 miles!

THE WHITE MOUNTAIN

Mauna Kea, the Hawaiian 'White Mountain', is one of the most massive natural structures in the world. It is what is known as a shield volcano, made up of numerous thin lava-flows built up one on top of the other. The lavas are so fluid that they can flow for great distances, producing a very broad base on the sea-bed.

Mauna Kea is one of five large Hawaiian volcanoes. The others are Kohala, Hululai, Mauna Loa and Kilauea. Of these the first two are extinct (at least, so one hopes!) and Hululai has not erupted since 1801, but Mauna Loa and Kilauea are active all the time.

The Earth's crust is divided into large 'plates' which move slowly around on top of the more viscous material below. When the idea of 'plate tectonics' was first proposed it was not taken seriously, possibly because its originator, Alfred Wegener, was a meteorologist rather than geologist, but it has now become fully accepted. The continents themselves are drifting around, and if one plate collides with another a mountain range is produced. Looking back far enough in time, most of today's continents were joined together in one gigantic land-mass; for example it is easy to see, as Wegener did, that the 'bulge' of South America fits neatly into the 'curve' of Africa.

A volcano is formed over a hot spot in the Earth's mantle, through which the magma can force its way to the surface. When the volcano drifts away from the hot spot it becomes extinct, and a fresh volcano is produced. In Hawaii, the extinct Kohala is at least 800,000 years old, and its plate has carried it away from the hot spot; the age of Mauna Kea is perhaps 600,000 years, but Mauna Loa and Kilauea are much younger, so that they are right over the hot spot at the present time.

The first white man to climb Mauna Kea was a botanist, Archibald Menzies, in 1794, but Big Island had been inhabited long before that, and Hawaiian mythology is very rich. Most of the legends feature Pele, the fire-goddess, who is not noted for her kindly nature. One story which always intrigues me is that of Pele and her sister, Poliahu the snow-goddess. It is said that at a great games festival held on the shore, by the slopes of Mauna Kea, Pele decided to join in, so she disguised herself as a beautiful woman and took part in activities such as surfing and spear-throwing. Unfortunately she realised that her sister Poliahu was even lovelier, and she became so angry that she summoned up her fire-forces; smoke and flames poured from the volcano's side, and the games broke up in disarray. Poliahu fled toward the summit, to find that even there the snows were being melted by Pele's wrath, but Poliahu was a match for her; drawing upon all her powers, she enveloped the whole of the summit with snow, so that the flames were extinguished, the pits and the cracks were sealed, and the remaining rivers of fire were quenched in the sea. Since then Pele has not returned to Mauna Kea, but on neighbouring Mauna Loa the battle between the two sisters continues, with volcanic eruptions bursting forth and snow falling every winter.

Perhaps it is appropriate that one of the most active volcanoes in the Solar System has been named 'Pele', but it is not on Earth; it is on Jupiter's satellite Io!

Marooned

There is no problem today in driving up to the summit of Mauna Kea from the half-way house, Hale Pohaku, but things were much less easy in the early days of the Observatory, when the road was very poor indeed. On one occasion a group of astronomers were marooned when the weather clamped down, and the road became impassable. The situation was really dangerous, both because of cold and because of the thinness of the atmosphere, and to keep warm they were reduced to burning some of the furniture. Luckily there is little fear of this happening again, because both road and vehicles have been so greatly improved, but it is true that the weather at the summit can be decidedly treacherous.

when it is active it can produce one of the most spectacular displays of pyrotechnics you can possibly imagine. The surrounding area has to be monitored all the time, and lava from Mauna Loa can flow almost as far as Hilo; it has been known to cut Saddle Road, which separates Mauna Loa from its neighbour Mauna Kea. Mauna Kea is extinct. It has not erupted for thousands of years, and will never do so again, which is why there have been no qualms about building a major observatory on top of it.

Though the summit of Mauna Kea is 13,796ft above sea-level, much of the volcano (a further 14,000ft) is beneath the waves, so that

as a structure it is considerably taller and more massive than Everest. The first man to consider building an observatory there was Gerard Kuiper, Dutch by birth but American by adoption. He knew the advantages of extreme altitude, and Mauna Kea beckoned him. His colleagues were less enthusiastic. As they pointed out, setting up equipment on or near the summit would be far from easy, because your lungs are taking in only about 39 per cent of your normal intake of oxygen and some people cannot tolerate this. Others are less affected by the conditions, and I have never been worried by them, partly perhaps because as a teenage flyer during the war I was used to operating at high altitude without an oxygen mask. Still, it must be said that it is not wise to go up to the summit and then start running up and down steps.

The recommended policy is to go first from sea-level to the 'half-way house', Hale Pohaku, at just below 10,000ft. Few people mind the rarefied air at this height, and it is the last 4,000ft which make all the difference, even though you can drive the distance in less than half an hour. Nobody sleeps at the summit, so that after a night's work the astronomers come down to Hale Pohaku. Moreover, oxygen is always available just in case anyone is taken ill.

Of the telescopes on Mauna Kea — and there are many — the one which has always intrigued me most is the UKIRT, or United Kingdom Infra-Red Telescope. (How nice to record that on Mauna

Opposite top
Summit of Mauna Kea

Opposite below
Cerro Tololo Observatory, also in the Atacama Desert; the main domes are fairly close together, rather than being spread out as at La Silla

Dome of UKIRT (United Kingdom Infra-red Telescope) on Mauna Kea, at sunset

Sacred caves

Kitt Peak, in Arizona, is now regarded as America's national observatory. It is 7,000ft above sea-level, and conditions there are excellent, so that planning began as early as 1950. There was, however, an initial problem: Kitt Peak comes in the reserve of the Papago Indians, and the sacred mountain Babuquivari is regarded by them as the centre of the universe, with the gods living in nearby caves. Negotiations were started, and were amicably concluded, with the Observatory acquiring the lease and promising that the sacred caves would never be disturbed. The first dome was erected in 1958, and today there are many more, plus the remarkable solar telescope which looks more like an inclined railway track than an astronomical instrument.

Fire at Palomar

Palomar, the site of the 200in Hale reflector, is a fascinating place. The lower slopes of the mountain are fertile enough; there are camping grounds, and also two reserves dedicated to the bald eagle and the spotted owl, both of which are officially classed as endangered species. It was therefore a most unpleasant shock when, on 29 July 1989, a serious fire broke out in the Dripping Springs camping ground, no doubt sparked off by some careless visitor.

Erratic winds soon made the fire spread, and before long it was out of control, despite the efforts of almost three thousand firemen who arrived from all over California. By 6 August the fire was within two and a half miles of the Observatory, and anything might have happened. Mercifully, the wind changed at the critical moment and by 9 August the threat to the Observatory was over, but over 15,000 acres had been

continued opposite:

Kea, at least, Britain is playing a major rôle.) The UKIRT has a 150in mirror which is very thin, and therefore cost less to build than a normal mirror of the same size would have done. The reason is that UKIRT, as its name implies, is designed specially for making studies in the infra-red part of the electromagnetic spectrum. Infra-red waves are much longer in wavelength than those of visible light, and therefore the accuracy of the telescope optics need not be so great. This was the reasoning, though in fact UKIRT turned out to be so much better than expected that it can be used for ordinary observation as well as in the infra-red.

The advantages of Mauna Kea are very obvious. It has even been said that from an astronomical point of view it is 'half-way to space'. It is doubtful whether any major observatory will be established at an altitude much greater than this, because it would be too risky, unless it could be operated entirely by remote control.

Until less than a century and a half ago, which is not long in the history of science, all that the astronomer could do was to look through his telescope and make drawings and notes, as some amateurs still do today. Then came photography, but the astronomer still had to be at his telescope; a time-exposure needs careful guiding, and in the pre-computer age no mechanical drive was accurate enough for more than a few consecutive minutes. During the 1920s and 1930s, for example, pioneers such as Edwin Hubble were using large telescopes to photograph spiral galaxies and other remote objects. To obtain a really good picture meant an exposure of many hours, and the Hubbles of the time had to be on the alert, checking constantly to make sure that the target had not drifted away from the field of view even by the slightest amount. However, in the 1970s, with the advent of modern-type computers and television techniques, the situation changed.

If you go to La Palma in the Canary Islands, and ascend the extinct volcano of Los Muchachos, you will find Britain's main observatory. There are two monster telescopes, the 102in Isaac Newton reflector, transferred there from its old site at Herstmonceux in Sussex, and the new 165in William Herschel Telescope or WHT. An observer using one of these telescopes will not be at the instrument itself, but will be sitting in a comfortable control room outside, drinking his coffee and studying the results coming through on a television screen. This has another advantage, too; temperature control is vitally important for telescope optics, and the warmth of an observer's body close to the telescope is no help at all.

We can take this a stage further. If the observer is not within touching distance of the telescope, why should he even be in the same building? No reason at all. With the aid of modern computers and electronics, he need not be in the same country. Today, the telescopes atop Mauna Kea can be moved around by observers in Britain, while some of the telescopes at La Silla in Chile can already be operated by observers at the German headquarters at Garching, near Munich. There may come a time when astronomers no longer have to do much travelling. They will be able to stay at home, and the observatories themselves will be manned only by maintenance staff, night assistants, and mechanics.

Some nations have no access to very high-altitude sites. Soviet Russia is one. In the 1970s the Russians built what was, and still is, the largest single-mirror telescope in the world. It has an aperture of 236in, so in theory it ought to collect much more light than its nearest rival in sheer size, the Palomar 200in. Unfortunately the site, in the Caucasus Mountains, is by no means ideal, and for this and other reasons the 236in has never really come up to expectations. Its main importance has been in the fact that it was the first giant telescope to have an altazimuth mounting, of the type since used on almost all major instruments — including the WHT on La Palma and the NTT in Chile.

There is also the question of latitude. Europe is unlucky inasmuch as so many of the most interesting objects in the stellar sky are too far south to be seen, whereas Australians and South Africans miss only the relatively barren areas near the celestial north pole; but it is a fact that early astronomy was northern-hemisphere astronomy, and the rich regions of the far south were perforce neglected. It is only in the second half of our own century that there has been a change of emphasis. Most of the great new observatories, apart from La Palma in the Canaries, are either in the southern hemisphere or else reasonably close to the equator.

There are no high mountains in Australia, and the Warrumbungle hills, where the Siding Spring Observatory has been built, are very modest; neither are the atmospheric conditions there anything like as consistently good as those at Chile, Hawaii or parts of the United States. Yet I would be prepared to claim that Siding Spring, with its AAT or Anglo-Australian Telescope, is the equal of any observatory in the world. It is so efficiently run, and is so accessible; you can drive there from Sydney in a few hours, and there is no need to make complex travel arrangements involving an aircraft which doesn't fly or a four-wheeled drive vehicle which breaks down. All you really need to do is to fit your car with 'roo bars', because kangaroos have absolutely no road sense. Siding Spring is also within striking distance of the great Parkes Radio Astronomy Observatory, which workers in both establishments find very useful indeed.

South Africa had problems of its own. There used to be several major observatories, but most of them were in areas which became more and more unsuitable because of the spread of cities such as Johannesburg and Pretoria. Eventually the South Africans took the plunge, and transferred almost all their main instruments to the best site in the Republic, Sutherland in Cape Province. It was a brave decision, and a good one, though the famous Radcliffe Observatory near Pretoria was a sad casualty.

Looking round the world, it seems that we now have a fairly effective spread of observatories, though some sites remain to be exploited — notably Antarctica. Yet the most unusual observatory is, without doubt, Homestake Mine. This time it is not a question of 'going up', but of 'going down'.

continued

devastated, plus three houses. The landscape 'looked like the bottom of a barbecue; charcoal-coloured, burned to a crisp and covered with white ash'. The total cost to the State was at least $8,000,000.

The Palomar Mountain General Store, halfway up the road (once owned by George Adamski, of flying saucer fame) promptly cashed in with T-shirts, each of which carried a picture of the Observatory with the design: FIRE OF '89. JUST WHEN YOU THOUGHT IT WAS SAFE TO GO BACK — PALOMAR II, THE SEQUEL. FIRE OF '89 BYOM. Most people had forgotten an earlier fire, in October 1987, which had devastated 16,000 acres. BYOM, by the way, means 'Bring Your Own Marshmallows'!

Lakeside observatory

The Yerkes Observatory, on the shores of Williams Bay not far from Chicago, is unusual for two main reasons. First, unlike most observatories, it is not high up, and actually adjoins a golf course. Secondly, the main telescope is not a reflector, but a 40in refractor, the first of the great telescopes masterminded by Hale. The funds were provided by the streetcar magnate Charles T. Yerkes, after whom the Observatory is named.

The Black Hills of South Dakota, outside Homestake Mine. The 'Black Hills' are usually covered with white snow!

South Dakota is the land of the gun-runners and the gold-miners. The gun-runners disappeared long ago, but the gold-mines are still there, and Homestake Mine, in the Black Hills, is one of the most important. Go down into the mine, and you will soon become pleasantly warm even though the temperature at the surface may be many degrees below zero. Go deep enough, and the rocks will be too hot to touch. I have never descended as far as that, because on my only visit to Homestake I left the cage at the first stop and made my way along the narrow tunnel to the Solar Observatory. It was quite an experience; one has to carry safety equipment to guard against a sudden surge of deadly carbon monoxide gas, and only a few inches above your head is a power cable which would fry you like an egg if you were unwise enough to touch it.

Here, a mile below ground, Dr Ray Davies and his colleagues of the Brookhaven National Laboratory have set up their observatory. Their telescope is simply a vast tank of cleaning fluid — thousands of gallons of it. The aim is to record strange particles from the Sun which are called neutrinos, and which are of immense importance in

Canopsus Road

When the Radcliffe Observatory was founded near Pretoria, after the war, local roads were named after stars: Rigel Road, Arcturus Road, Sirius Road, and so on. One of these was named Canopsus Road. Jack Bennett, a well-known amateur astronomer who lived in Pretoria, wrote to the City Council and gently pointed out that it really should be *Canopus,* not Canopsus. Their reply is worth putting on record: 'Thank you for your letter. Unfortunately our maps have now been printed and distributed, and it is not possible to alter the name of the road. Can you not alter the name of the star?'

The Radcliffe Observatory has been dismantled, and the telescopes transferred to Sutherland in Cape Province, but when I was last in Pretoria, a few months ago, Canopsus Road was still Canopsus Road!

The 60in reflector at Mount Wilson – the first really large telescope on the mountain. It is still in full use

Kitt Peak, Arizona, the American observatory which has as its main instruments the solar telescope and the 158in reflector, here shown in the dome to the left
(Peter Gill)

any study of solar physics. Neutrinos are very odd things, because they have no electrical charge and apparently no mass, so that they can pass unchecked right through the Earth (we are being bombarded by them at the present moment). But they can sometimes be trapped by interacting with atoms of chlorine, turning the chlorine into a gas known as argon, which in this form is mildly radioactive. Cleaning fluid contains a great deal of chlorine, so the idea is to flush out the tank periodically and see how many atoms of chlorine have been changed into argon. This will show how many neutrinos have scored direct hits.

It is essential to go down a mile below ground level, as otherwise different particles coming from space — cosmic rays — would produce the same effects and ruin the analyses. But cosmic rays cannot penetrate deep below the Earth's surface, whereas neutrinos have no trouble at all. This is why the Observatory has been built deep in the gold-mine. It has been operating for many years, and it works well, though it is providing information which has baffled theoretical astronomers more than some of them care to admit.

Yes: astronomers go to strange haunts. In their pursuit of the stars, they are as ready to scale mountains as they are to burrow, mole-like, into the Earth's crust. Some astronomers have already travelled into space, and during the coming century we will undoubtedly set up observatories on the surface of the Moon. At least one has to admit that astronomers are nothing if not adventurous!

GREAT TELESCOPES

Some of the world's great telescopes are:

REFLECTORS

Observatory	Aperture (in)
Hawaii (Keck Telescope)	396
Zelenchukskaya, Russia	236
Palomar, California (Hale Reflector)	200
La Palma, Canary Islands (William Herschel Reflector)	165
Cerro Tololo, Chile	158
Kitt Peak, Arizona	158
Siding Spring, Australia (Anglo-Australian Telescope)	153
La Silla, Chile	158
La Silla, Chile (NTT)	158
Mount Stromlo, Canberra, Australia	150
Mauna Kea, Hawaii (UKIRT)	150
Mauna Kea, Hawaii (Canada-France Hawaii Telescope	143
Lick Observatory, California	119
Mauna Kea, Hawaii (NASA Infra-red Telescope)	119
McDonald Observatory, Texas	106
Crimean Astrophysical Observatory	103
La Palma, Canary Islands (Isaac Newton Telescope)	100
Las Campanas, Chile (Irénée du Pont Telescope)	100
La Palma, Canary Islands (Nordic Telescope)	100
Space Telescope (HST)	94

(Of the Chilean stations, Cerro-Tololo and Las Campanas are run from the United States, La Silla by the European Southern Observatory. The WHT and the INT on Canary Islands telescopes are run by Britain, while the new Nordic telescope is due to Norway, Sweden, Denmark and Finland together.)

REFRACTORS

Observatory	Aperture (in)
Yerkes, Willams Bay, USA	40
Lick, California	36
Meudon, Paris, France	32.7
Potsdam, Germany	32
Allegheny, Pittsburgh	30
Nice, France	30

(The Pittsburgh telescope was built in 1914; all the other refractors in the list are pre-1900.)

The Rosse 72in reflector at Birr Castle

Most people picture an observatory as being a graceful dome, housing a large telescope. In many cases this is true enough, but early observatories were quite different, and until the seventeenth century they did not even have telescopes. Observers such as Tycho Brahe had to make do with measuring instruments used with the naked eye.

Most of the really large telescopes made before 1900 were refractors, though there were some notable exceptions; William Herschel made excellent reflectors, and there was also the unique 72in reflector made and used by the Earl of Rosse in Ireland. But these mirrors were of metal; only when it became possible to make glass mirrors, covering them with a thin layer of material such as silver or aluminium, did the reflector really come into its own.

An object-glass for a refractor has to be supported all round its edge, and if it is too heavy it will distort under its own weight, making it useless. The largest of all refractors is that at the Yerkes Observatory, near Chicago. Its 40in object-glass is not likely to be surpassed in size; a 49in was once cast in France, but proved to be a total failure.

With a reflector, the mirror can be supported on its back, because the light does not have to pass through it, and in any case a large mirror is much easier to make than a large lens. In 1908 George Ellery Hale, an American astronomer, who had a remarkable knack of persuading friendly millionaires to finance his schemes, was responsible for setting up a 60in reflector on the summit of Mount Wilson in California; this was followed in 1917 by the Mount Wilson 100in, and then, in 1948, by the 200in reflector on Palomar Mountain, also in California. Only one larger single-mirror telescope has been built since: the 236in in Soviet Russia, which, to be candid, has never been a real success, partly because of technical problems with the telescope itself and partly because of the atmospheric conditions in the Caucasus, where it has been set up.

Atmospheric conditions are all-important. The Earth's air is the chief enemy of the astronomer, because it is dirty and unsteady as well as blocking out many of the incoming radiation; light pollution is another problem, and in fact there was a period when the Mount Wilson 100in was 'mothballed' because of the increasing glare from Los Angeles.

Not all present-day telescopes are of the single-mirror type. One interesting experiment was the MMT or Multiple-Mirror Telescope; this used six 72in mirrors working together, making an instrument equivalent to a single 176in. It worked well, but is now being dismantled, and the six mirrors will be replaced by a single much larger mirror made by Roger Angel using his new 'spin-honeycomb' technique. The Keck Telescope on Mauna Kea is the first really large telescope to have a segmented mirror; the 36 separate segments make up a mirror equivalent to a single 396-inch. Next will come the VLT or Very Large Telescope, consisting of four 8-metre telescopes working together. When complete, soon after 2000, this will be the most powerful optical telescope in the world. Even La Silla was not considered a good enough site; the VLT will go to Cerro Paranal, in the northern Atacama Desert, where rainfall is unknown. Work started in 1993, and the complete top of a graceful mountain was sliced off and flattened!

Radio astronomers have different needs; they are not bothered by clouds or light pollution, but they are very worried indeed by commercial and military radio interference, which is becoming increasingly obtrusive. (Sir Bernard Lovell once commented that unless something was done, it could well be that radio astronomy from the Earth's surface would be a science limited entirely to the second half of the twentieth century.) The most famous of all radio observatories is at Jodrell Bank in Cheshire, with the 250ft 'dish' of the Lovell Telescope, though the steerable 'dish' at Bonn in Germany is larger. At Puerto Rico there is the Arecibo radio telescope, 1,000ft across, built in a natural hollow in the ground. Obviously it cannot be steered, but by ingenious techniques it can be used to cover most of the sky.

The most controversial of all telescopes has been the Hubble Space Telescope, a 94in reflector which was launched into space by the shuttle in 1990. Moving round the Earth at a height of 300 miles, above virtually all of the atmosphere, it was designed to 'see' more clearly than any telescope could do from the Earth's surface. Unfortunately, it was found – too late! – that the main mirror had been incorrectly figured. It would be quite wrong to dismiss the telescope as a failure, and in many ways it can still out-perform any Earth-based telescope, though some of its planned programmes have had to be modified.

In the future, of course, there may well be an observatory on the lunar surface. Radio astronomers, in particular, are thinking ahead to a station on the Moon's far side, which is completely 'radio quiet'.

Dome of the new 100in Nordic telescope at La Palma. This – the world's newest large reflector – is a joint project by Norway, Sweden, Denmark and Finland

The Lives of the Stars

The Herzsprung-Russell Diagram

These diagrams, known as HR Diagrams for short, are of fundamental importance in all studies of stellar evolution. In them, the stars are plotted according to their spectral types and their luminosities. It is obvious at first glance that most of the stars lie on a line extending from the top left (very hot, powerful white or bluish stars) down to the lower right (very dim red stars); this is called the Main Sequence, with the Sun as a typical Main Sequence star. The giant branch lies to the upper right. The division into giants and dwarfs is also very obvious (the white dwarfs, to the lower left in the Diagram, come into an entirely different category). It was natural to assume that a star began its career as a cool giant, moved across the Diagram to the upper left, and then slid down the Main Sequence to end up as a faint red dwarf. This was later found to be very wide of the mark, though it is quite true that all stars begin their life-stories by condensing out of the huge, tenuous gas-and-dust clouds which we call nebulæ.

One aspect of astronomy which strikes most people is its apparent timelessness. There are things which happen quickly; a meteor may flash across the sky, a bright aurora may shine down with its flickering, vibrating colours, or a star may suddenly be blotted out by the advancing body of the Moon, but generally speaking everything happens at a leisurely pace. Even the daily rotation of the sky is not perceptible except over periods of minutes.

What, then, about our Sun? Is it eternal, or will it change over the ages? So far as we are concerned, it is always much the same, but we know that it is anything but eternal. Several thousands of millions of years ago it was born inside a nebula, and several thousands of millions of years hence it will swell out to become a terrifying red giant star, with fatal consequences to the Earth unless we have by then learned enough to tamper with Nature itself.

Sun worship goes back a long way, and this is understandable, because we depend so completely upon what the Sun sends us. There was even a religion based solely upon it. Between 1387 and 1366BC Amenophis IV, Pharaoh of Egypt, took a fundamental step; he changed his name to Akhenaten, built a new capital city, and introduced a code which was different from anything before or since, if only because it was based upon something a great deal more telling than dogma. Read Akhenaten's *Hymn to the Sun,* and you will probably see what I mean. Predictably, the orthodox religions managed to destroy Akhenaten's work within months of his death, and the Sun became once more subordinate to deities of various degrees of ferocity. Looking at the religious wars which have gone on unceasingly ever since, and which still go on today, one feels that we could learn a great deal from Akhenaten.

The Sun itself looks placid enough, but, as I have said, it is dangerous even when it is shining through a thick layer of haze or mist and seems deceptively innocent. It is, after all, a huge hydrogen bomb, albeit a controlled one. It is hard to realise that it is nothing more than a dwarf star.

The Sun undoubtedly shows minor fluctuations over long periods. Throughout the Earth's history, which dates back some 4,700 million years or so, there have been periodical Ice Ages, the last of which ended a mere 10,000 years ago, and there is evidence that these cold spells have been due to variations in the Sun's output — though by no means all authorities agree about this, and the Ice Ages may well be much more complicated in origin. Yet over a lifetime, or even over the whole story of civilisation, we have been unable to see the Sun 'age', and the same is true of the stars; again we come back to the leisurely pace of events in the universe. So in trying to trace the life-story of a star, our only method is to identify different stars at different stages of their evolution, and then try to put the story together.

I have compared this situation with that of a visitor from Outer Space who arrives on Earth without having any preconceived ideas about *Homo sapiens*, and is allowed to spend half an hour in Trafalgar Square. He will see babies, boys, youths, men and old men, so that if he is intelligent he will be able to trace the evolution of a male human being. Of course, he will not be able to find out where the baby came from (unless someone has been kind enough to acquaint him with the facts of life), but at least he will be able to form an overall picture — provided that he can tell which human beings are old and which are young.

We have the same sort of problem with the stars. The key here lies in colour. We have stars which are bluish, while others are white, yellow, orange or red, and our first task is to decide which are

Unwise prediction

August Comte, a French philosopher who lived from 1798 to 1857, was the inventor of the term 'sociology', and according to one reference book 'advocated a positive social organisation with a rational religion somewhat in the style of the Enlightenment and the French Revolution'. In 1835 he stated that there were some areas of knowledge which would be forever hidden from mankind, and he gave as an example the chemistry of the stars. All of which goes to show that when a French philosopher makes a profound statement, he is almost certain to be wrong.

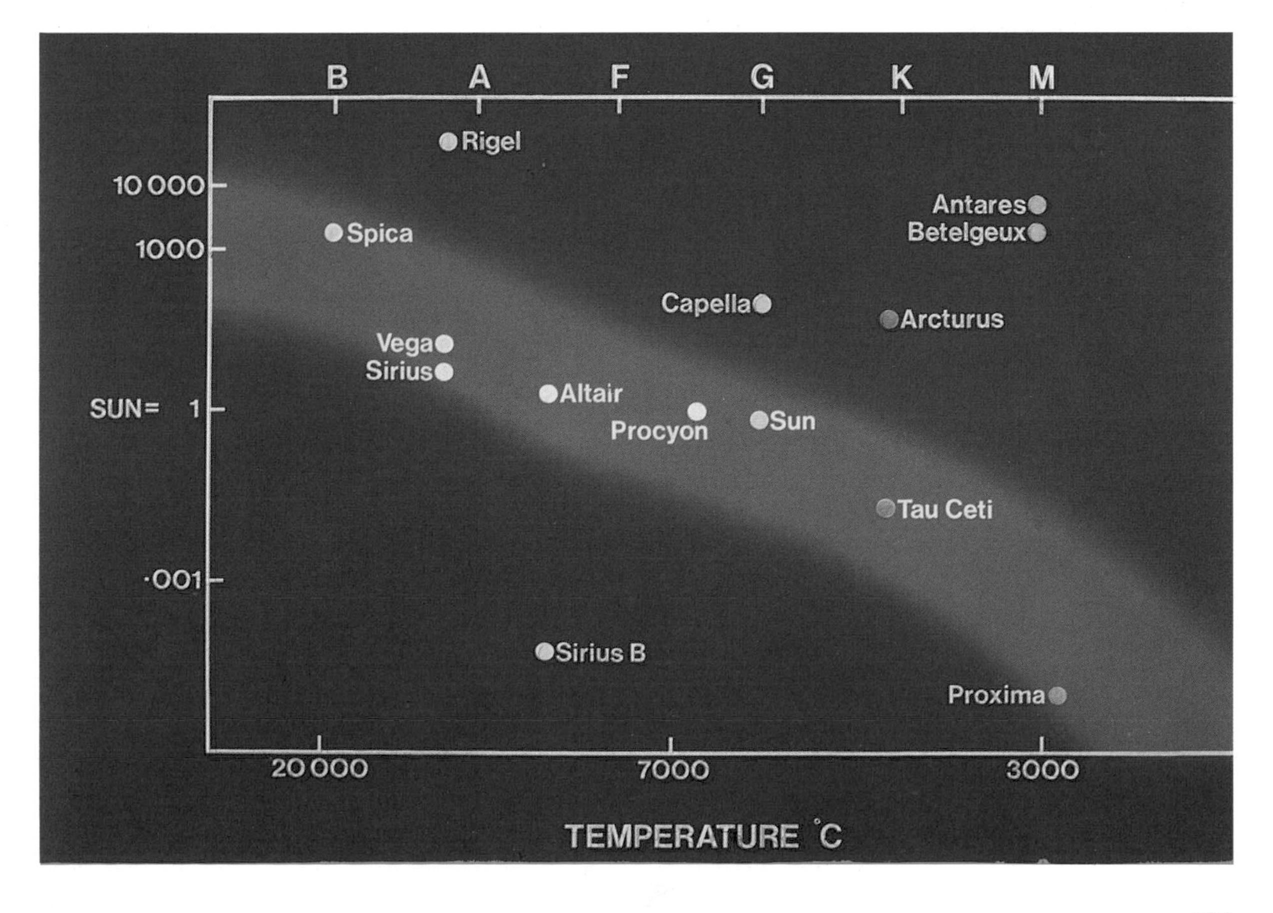

The Hertzsprung-Russell Diagram, with various stars plotted

Red Sirius?

Sirius is a pure white star, but many ancient writers recorded that it was red — among them Virgil (in the Aeneid), the Roman historian Seneca, and Ptolemy, the last of the great astronomers of Classical times, who died around AD180. Considerable evidence has been collected, but there seems no chance that Sirius itself has changed; it is simply not that sort of star, and is stable on the Main Sequence. True, it twinkles strongly, but there is no way in which it could now be described as red.

Various theories have been proposed, one — due to Sir John Herschel — being that Sirius was temporarily reddened by an interstellar cloud passing in front of it; but most of the suggestions involve the Companion, which must have passed through a red giant stage before collapsing into the white dwarf condition. However, the time-scale is all wrong, and in any case the present Sirius, shining together with an even more luminous red giant, would produce a 'star' as brilliant as Venus, and of this there is no hint. Presumably there is an error in interpretation somewhere or other; all the same, it is decidedly curious.

youthful and which are senile. Until only about half a century ago, astronomers picked wrong. Standing under a starlit sky, it is easy to appreciate the difficulty; if you look at Orion, which star is the younger — the glittering white Rigel or the orange-red Betelgeux? And where does our yellow Sun fit into the story?

Even before the 'modern' way of thinking, it was generally agreed that nebulæ were stellar nurseries. It seemed logical to assume that a star condensed, by gravitation, out of the gas and dust spread through space. As it became denser, the fledgling star would also become hotter near its centre, so that eventually it would start to shine. At first it would be unstable, and would flicker irregularly, but then it would 'settle down' and become more stable.

Astronomers were in general agreement about this, but before going any further it was vital to decide how a star produces its energy. Simple 'burning' would not do, because it would not produce enough energy for a sufficient length of time. Another idea was to assume that a star produced its energy by slow gravitational shrinking, and this was the basis of early theories of stellar evolution. By 'early' I mean between around 1912, when Ejnar Hertzsprung of Denmark and Henry Norris Russell of the United States drew up their famous Diagram, and around 1940, when all our ideas began to change.

Hertzsprung and Russell looked at stars of different kinds, and found that there were two completely different classes of red, orange, and to a lesser extent yellow stars: giants and dwarfs. Red stars were either very luminous (such as Betelgeux in Orion) or else very feeble (such as Proxima Centauri, the nearest of all stars beyond the Sun). Red stars about the same power as our Sun did not seem to exist at all. From this, an evolutionary sequence was worked out. A star would condense out of a nebula as a large, cool globe, glowing redly; it would condense and heat up, becoming first an orange giant, then a yellow giant and then an energetic white star, after which it would fade gradually away, changing colour through orange and yellow before becoming a dim, exhausted red dwarf. Finally it would end up as a cold, dead globe — a black dwarf.

It all seemed straightforward enough, but from the outset there were nagging doubts about it, mainly because of the time-scale. A star simply would not last for long enough to satisfy the theorists. Remember, the Earth was known to be well over 4,000 million years old, and the Sun was inevitably older than that, so that some extra energy source was needed. It was suggested that there was an actual 'annihilation of matter' going on deep in a star's core, and in this case a star could go on shining for millions of millions of years, but this was just as obviously too long as earlier theories had been too modest. The clue to the whole problem was given in the early years of World War II by two astronomers working quite independently, Hans Bethe in America and George Gamow in Russia.

Bethe and Gamow realised that nuclear reactions involving the transformation of hydrogen into helium were responsible for the energy production of normal stars such as the Sun. The process is rather complex, but it works, and it means that the evolution of a star depends mainly upon its initial mass. The more massive the star, the more quickly it runs through its career.

In each case, the star begins by condensing out of the material in a nebula. If the original mass is very low by stellar standards, nuclear reactions will never be triggered off, and the result will be a body which is 'not quite a star and not quite a planet' — a sort of missing link, known misleadingly as a brown dwarf. Nothing more will happen, and eventually the failed star will become cold and dead.

If the mass is greater, the core temperature will reach the critical value of about 10,000,000°C, and nuclear reactions will begin, so that the star will join the Main Sequence of the Hertzsprung-Russell Diagram. This is what the Sun has done; it is about midway through its Main Sequence stage. Yet the hydrogen 'fuel' cannot last for ever, and when it runs low the Sun will have to change drastically. Different sets of reactions will take over, so that the core will shrink and heat up while the outer layers will expand and cool down. The Sun will leave the Main Sequence, and will turn into a red giant, as Betelgeux in Orion is now.

This will spell disaster for the Earth, because we cannot hope to survive the blast of radiation from a swollen Sun emitting at least a hundred times as much energy as it does now. But the red giant stage is temporary; after a while the outer layer will be thrown off altogether, and all that will be left will be a Sun stripped of its surroundings. It will have turned into a white dwarf. And to introduce white dwarfs, let us turn to Sirius, the most brilliant star in the sky, which has a white dwarf companion.

You cannot mistake Sirius whenever it is above the horizon,

Double stars

In his book *The Sirius Mystery*, published in 1975, R.K. Temple claims that an African tribe — the Dogon, in Mali — knew about the Companion of Sirius long before modern astronomers found it. This means that either they had telescopes (!) or that they were drawing upon Ancient Teachings. Neither of these ideas seems very likely, and in fact Dogon religion makes virtually everything double, so that Sirius would be no exception.

The Milky Way, showing Sirius at the upper right

Companion of Sirius

When the companion of Sirius was discovered, it was naturally assumed to be large, cool and red. In 1907 the well-known Irish astronomer J.Ellard Gore wrote as follows:

> If its faintness were merely due to its small size, its surface luminosity being equal to that of our Sun, the sun's diameter would be the square root of 1,000 or 31½ times the diameter of the faint star, in order to produce the observed difference in light. But on this hypothesis the Sun would have a volume 31,500 times the volume of the star and, as the density of a body is inversely proportional to its volume, we should have the density of the Sirian satellite over 44,000 times that of water. This of course is entirely out of the question.

One cannot blame Gore; the idea of a star as dense as this seemed to be outrageous — but it is true; the 'Pup' is only about 24,000 miles across, and 60,000 times as dense as water. Gore, alas, never knew; at the age of only fifty-five he was killed in a street accident in Dublin.

because it is so outstanding; it far outshines any other star of the night sky, and it dominates the evening sky all through the late winter and early spring (in the northern hemisphere of the Earth, that is to say; in the southern hemisphere, of course, the seasons are reversed). In case of doubt, simply follow the line of the three stars making up Orion's Belt. Sirius is the leader of the constellation of Canis Major, the Great Dog, and so it is often called the Dog Star. To the ancient Egyptians it was of special importance, because they used it to regulate their calendar. When they could first see Sirius in the dawn sky, just before sunrise, they knew that the Nile was about to flood, and this was vital to their whole economy.

Sirius is one of our nearest neighbours. It is only 8.6 light-years away, which corresponds to roughly 50 million million miles, and of all the naked-eye stars only Alpha Centauri in the far south is closer than that. Sirius is hot, with a surface temperature of over 10,000°C against less than 6,000°C for our yellow Sun, and it is moderately luminous by stellar standards; it is the equal of 26 Suns. In every way it is a typical Main Sequence star.

Because it is so near, relatively speaking, it has a measurable individual or 'proper' motion against the background of more remote stars. During the last century it was found that this motion is not regular; Sirius seems to be 'wobbling' slightly as it moves. Friedrich Wilhelm Bessel, the German astronomer who was the first to measure the distance of a star, correctly claimed that this slight, slow wobbling was due to an invisible companion star which was pulling the bright Sirius out of position, and in 1862 the companion body was duly discovered, almost exactly where Bessel had said that it ought to be. It was faint, with only 1/10,000 the luminosity of Sirius itself, but it was quite unmistakable. The way in which it moved gave the key to its mass, which turned out to be about the same as that of the Sun. It and the bright star were moving round their common centre of gravity in a period of fifty years.

There was nothing surprising about his, because binary pairs are very common indeed. The companion star — often nicknamed the Pup — was assumed to be large, cool and red, but in 1915 W.S. Adams, at Mount Wilson, was able to study its spectrum, and he had a tremendous shock. The Pup was not red; it was white, and was very hot indeed.

Astronomers were puzzled. If the Pup were hot, it would have to be very small, as otherwise it would look much brighter in the sky than it actually does. Yet the mass was certainly equal to that of the Sun, and it followed that the density would have to be thousands of times greater than that of water, which did not seem very likely.

Further observations were made, with the same result, and astronomers had to accept the existence of a star which 'weighed' at least 60,000 times as much as an equal volume of water would have done. Sirius B, to give it its official title, was the first known white dwarf. Other white dwarfs were soon identified, some of them even more extreme, and it became clear that they are very common in the Galaxy; they had not been detected earlier because they are too dim

Modern photograph of the Crab Nebula: the structure is very evident

to be seen at all unless they are reasonably close to us by cosmical standards.

As soon as the source of stellar energy had been worked out, which, as we have seen, was not until a quarter of a century after the revelations concerning the Pup, everything began to fit neatly into a general scheme. When a solar-type star has passed through its red giant stage, the atoms making up the old core are broken up and packed tightly together with almost no waste of space, which accounts for the high density. A white dwarf is a bankrupt star; it has no further reserves of nuclear energy upon which it can call, and eventually it must lose the last of its light and warmth, though the process is bound to take so long that we cannot even be sure that the universe is old enough for any of these dead 'black dwarfs' to have appeared.

If the initial mass of a star is more than about 1.4 times that of the Sun, everything happens at an accelerated pace, and the last part of the story is completely different. After the star has used up its available 'fuel', energy production suddenly stops. There is a violent collapse, followed by a shock-wave which blows the main star to pieces in what is termed a supernova outburst. The component parts of the atoms in the core are compressed to a much greater degree even than in a white dwarf, and we are left with a neutron star with a density thousands of millions of times greater than that of water. The neutron star will spin round rapidly, and it will have an intense magnetic field. It will also send out pulsed radio waves, which is why neutron stars are often called 'pulsars'.

Pulsars, too, are bankrupt; over the ages their spin rate will slow down, and eventually they will cease to radiate at any wavelength. No doubt many dead pulsars wander through the Galaxy, though we have no way of finding them, and in fact very few have been optically

THE SPECTRA OF THE STARS

It was Isaac Newton who first realised that 'white' light is really made up of all the colours of the rainbow, from red through to violet. Passing sunlight through a glass prism, he produced the first 'spectrum', though he never took his experiments much further — possibly because the only prisms available to him were of poor-quality glass.

In 1814 Josef Fraunhofer made the first really scientific study of the Sun's spectrum, showing that it was made up of a rainbow background crossed by dark lines which are still often called the Fraunhofer lines. These lines were always in the same positions, and of the same intensities, but it was left to Gustav Kirchhoff, more than forty years later, to find out how they are caused.

Kirchhoff showed that spectra are of two distinct types. Analyse the light from a luminous solid, liquid, or gas at high pressure, and you will have a continuous or rainbow spectrum from red to violet. However, when the light from an incandescent low-pressure gas is analysed the result will not be a rainbow; instead, there will be disconnected bright or emission lines. Each bright line is the special trademark of one particular substance. For example, two bright yellow lines must be due to sodium, one of the two elements making up common salt; nothing else can produce them.

Now consider the spectrum of a star. The bright surface is gas at relatively high pressure, and so yields a continuous or rainbow spectrum. Above it is a layer of gas at lower pressure. Normally this layer would produce a bright-line or emission spectrum, but because of the rainbow background the lines are 'reversed', and appear dark, giving what is termed an absorption spectrum. The positions and intensities of the lines do not change, and so they can be identified. To give just one example: two prominent dark lines in the yellow part of the rainbow correspond to the two bright yellow lines of sodium, telling us that there must be sodium in the star.

By studying the spectra of the stars we can therefore find out which elements are present there. For convenience, the stars have been divided into various spectral types:

Type	**Colour**	**Approximate surface temperature (°C)**	**Example**
W	White or bluish	Up to 80,0000	
O	White or bluish	35,000	
B	Bluish white	25,000	Rigel
A	White	10,000	Sirius
F	Slightly yellowish	7,500	Polaris
G	Yellow	6,000	Capella, the Sun
K	Orange	4,000	Arcturus
M	Orange-red	3,400	Betelgeux
R	Reddish	2,600	
N	Reddish	2,500	
S	Reddish	2,500	

You may wonder why the sequence does not run A,B,C... Originally it was meant to do so, but various modifications had to be introduced until the result was alphabetically chaotic. You may care to remember it by the celebrated mnemonic Wow! O Be A Fine Girl Kiss Me Right Now Sweetie (though according to some distinguished astronomers, S should really stand for Smack!).

Most of the stars belong to types B to M. The sequence denotes decreasing surface temperature, though it is not an indication of evolution as used to be thought. Note also that types R and N are now often combined into a single type, C.

Supernova in external galaxy (the supernova is arrowed)

identified, though there is a pulsar in the famous Crab Nebula and another in the southern constellation of Vela, the Sails. But suppose we have a star which is more massive still — say ten times as massive as the Sun? Again there is a different end to the story. Once the grand collapse starts, there is no way of stopping it. Gravitation takes over, and the old star simply goes on shrinking without even having the chance to explode as a supernova. As it contracts, its density increases at an amazing rate, and the escape velocity goes up. The end product is neither a white dwarf nor a pulsar; it is a black hole.

Escape velocity is a term which has become widely used in everyday language since the start of the Space Age. To break free from the Earth, you must work up to a starting velocity of 7 miles per second. The escape velocity of the less massive Moon is a mere 1½ miles per second, which, as we have seen, is the reason why the Moon has no atmosphere; its pull was too feeble to hold down any air there may have once been. The Sun's escape velocity is 465 miles per second. With a very massive collapsing star, the escape velocity finally reaches 186,000 miles per second, which is the velocity of light — and light is the fastest thing in the universe. If light cannot escape from the collapsed star, then certainly nothing else can do so. The dying star is surrounded by a 'forbidden region' which is to all intents and purposes cut off from the rest of the universe, and this is what we mean by a black hole.

We cannot see a black hole, because it emits no radiation at all, but

View from the bottom of a mine

There is an old story that if you go down to the bottom of a mine, and look straight up, you will see stars even in the daytime. Actually this is not true, because what matters is the contrast between the brightness of the star and the brightness of the sky — and going down a mine-shaft does not make the slightest difference. Apart from the Sun and the Moon, the only object which can be seen with the naked eye in broad daylight is Venus, and then only by people with exceptionally sharp sight.

its gravity can still affect any object within its range, and this is how we believe that black holes can be tracked down.

We cannot be certain about this, and it is fair to say that some astronomers are still sceptical about the whole black hole concept, but on the whole it seems that they probably do exist, even though a black hole must be the most bizarre thing that we can imagine. Inside the 'event horizon', the name given to the boundary of the black hole, all the ordinary laws of science break down. For example, what happens to the remnant of the collapsed star? Does it go on shrinking until it has no volume at all, so that it literally crushes itself out of existence? We do not know. There have been wild ideas about the possibility of plunging into a black hole and emerging in a different part of the universe, or in a different universe altogether, but I admit that personally I take these theories with a very large grain of cosmic salt, and certainly there is no way of visiting a black hole to find out!

One thing is definite: our Sun is not massive enough to explode as a supernova, and not nearly massive enough to become a black hole. It will undoubtedly become a red giant, and this is a crisis which our distant descendants will have to face, assuming of course that mankind still exists on Earth at that remote epoch.

There will be plenty of warning. In around 2,000 million years from now there will be a perceptible change of climate as the Sun grows more luminous; the polar ices will melt, sea-levels will rise, and our present frigid zones will become Mediterranean. Gradually,

The supernova SN 1986A in the Large Cloud of Magellan, near maximum. The red mass to the upper left is the Tarantula Nebula: the supernova is to the lower right of this nebula (South African Astronomical Observatory)

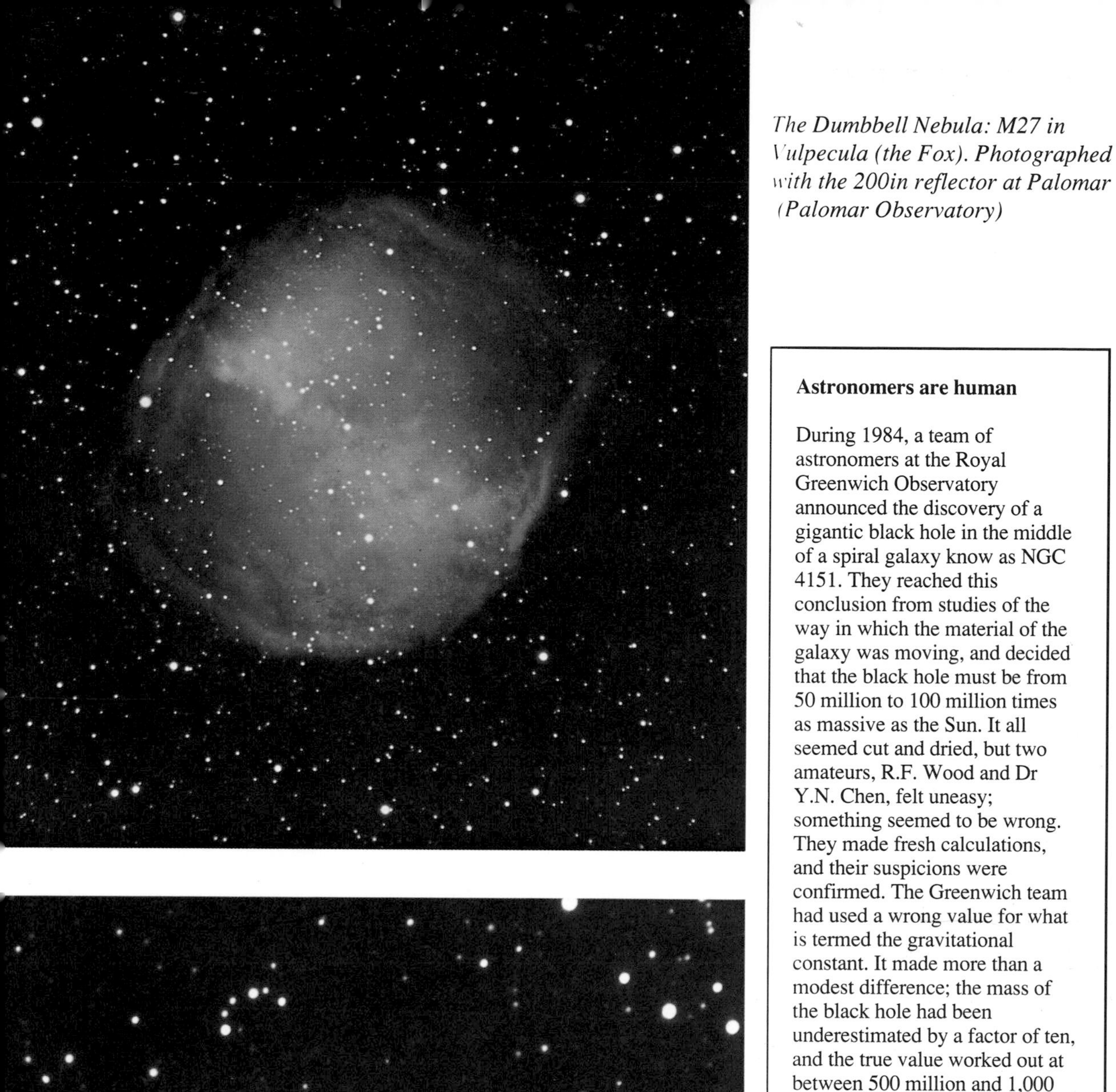

The Dumbbell Nebula: M27 in Vulpecula (the Fox). Photographed with the 200in reflector at Palomar (Palomar Observatory)

Astronomers are human

During 1984, a team of astronomers at the Royal Greenwich Observatory announced the discovery of a gigantic black hole in the middle of a spiral galaxy know as NGC 4151. They reached this conclusion from studies of the way in which the material of the galaxy was moving, and decided that the black hole must be from 50 million to 100 million times as massive as the Sun. It all seemed cut and dried, but two amateurs, R.F. Wood and Dr Y.N. Chen, felt uneasy; something seemed to be wrong. They made fresh calculations, and their suspicions were confirmed. The Greenwich team had used a wrong value for what is termed the gravitational constant. It made more than a modest difference; the mass of the black hole had been underestimated by a factor of ten, and the true value worked out at between 500 million and 1,000 million times that of the Sun.

Rather diffidently, the amateurs sent in their results. The Greenwich team promptly and generously admitted the error; somehow or other it had slipped through unnoticed not only by the original authors, but also by the other equally eminent professionals who had refereed the paper.

At least the episode shows that astronomers are human!

Planetary nebula in Aquila (the Eagle). Photographed with the 48in Schmidt telescope at Palomar (Palomar Observatory)

but inexorably, the trend will continue, until the Sun becomes a fully fledged red giant. By then Mercury and Venus will have been destroyed, and on Earth the surface temperature will rise to over 1,500°C; the daytime Sun will almost fill the sky, but there will be nobody here to see, because the whole world will have been transformed into an ocean of molten lava. Eventually the atmosphere will be stripped away, and at the peak of the Sun's dying spasm it is more than likely that the Earth will be vaporised altogether. The more

PLANETARY NEBULÆ

Some names in astronomy are frankly misleading. The Straight Wall on the surface of the Moon is not straight, and is not a wall; though 'nova' means 'new' in Latin, a nova is not a new star; and an eclipse of the Sun is not really an eclipse, but an occultation of the Sun by the Moon. Even worse is the term applied to the lovely though faint objects known as planetary nebulæ. They have absolutely nothing to do with planets, and in the proper sense of the term are equally unrelated to nebulæ. The name is probably due to the French astronomer Antoine Darquier, who discovered the best-known member of the class, the 'Ring Nebula' in Lyra, in 1779, and described it as 'a perfectly circular disk as large as Jupiter, but dull and looking like a fading planet'.

The Ring Nebula, not far from Vega in the sky, is well below naked-eye visibility, and I have never been able to see it with binoculars (though some people tell me that they can do so). A small telescope will show it in the form described by Darquier, and greater power reveals that it looks like a tiny, luminous cycle-tyre. There is a star right in the middle of the ring, but to see it you need a fairly powerful instrument. The Ring Nebula is easy to locate, as it lies directly between the two naked-eye stars Beta and Gamma Lyræ.

Over a thousand planetary nebulæ are known in the Galaxy, and the total number is probably at least ten times as great. They are stars in the transition stage from middle age to senility, so that they have already left the giant branch (upper right of the HR Diagram) and have thrown off their outer atmospheres; the central stars are very hot, and are well on the way toward becoming white dwarfs. The Ring Nebula is 1,400 light-years away, and its diameter is roughly half a light-year (that is to say, 30,000 times the distance between the Earth and the Sun); most of the gas is hydrogen, with an admixture of helium, while the central star has a surface temperature of around 100,000°C, and is several thousand times denser than the Sun.

But is the 'ring' really a ring? Can it be that we are seeing what is in reality a shell of gas, and that the ring aspect is due to the greater amount of material lying in our line of sight when we look at the edge of the shell? At any rate, not all planetaries are symmetrical. The Dumbbell Nebula in the constellation of the Fox lives up to its nickname, and so does the Owl Nebula in the Great Bear, with its two 'eyes'; others are referred to as the Helix or Sunflower, the Butterfly and (with less obvious reason) the Eskimo.

All planetaries are expanding, as we can tell from spectroscopic observations as well as direct viewing. If the Ring Nebula has been expanding at its present rate ever since its formation, its age *as* a planetary can be no more than 20,000 years, which is reasonable enough. Whether the process of formation is the same in all cases we do not know; presumably a single 'puff' would result in a symmetrical object such as the Ring, while with less regular planetaries there may have been a whole series of outbursts. In any case, the end result is the same: the outer layers of the old giant star are stripped away, leaving only the small, hot core.

The central stars must fade with time, and so the nebulosity must fade too, becoming larger and more diffuse until it ceases to shine, leaving the old star as a true white dwarf. There are, too, a few cases of 'protoplanetary nebulæ', one of which, FG Sagittæ in the constellation of the Arrow, seems to have been caught in the act, because it brightened steadily in the years following 1894 and has now become fuzzy, so that it is sending out a second 'shell'. In a few centuries it may well have turned into a recognisable planetary nebula — one of the rare instances of our being able to see a star change state as it grows old.

There are several planetaries within range of modest telescopes, and they are always interesting to find. In the far future our Sun will itself pass through the planetary nebula stage, though we will not be here to see it.

distant planets will survive, but when the Sun becomes a white dwarf, with only a thousandth of its present luminosity, the Solar System will be a very gloomy place indeed. All that will be left will be a tiny, feeble Sun still being circled by the ghosts of its remaining planets.

But for the present, and for the foreseeable future, we are safe from the Sun. Ice Ages may come and go, but so far as we are concerned the Earth will remain very much as it is now. The only danger to life on our world at present comes not from the Sun, but from ourselves.

SUPERNOVÆ

Supernovæ are the most colossal outbursts in all Nature. Many of them have been discovered in outer galaxies, but in our own system only four have been definitely seen during the last thousand years: those of 1006, 1054, 1572 and 1604, all of which became brilliant enough to be seen with the naked eye in broad daylight. Since then the only supernova to have become bright enough to be seen without optical aid was that of 1987, which burst forth in the Large Cloud of Magellan, 170,000 light-years away.

They are of two distinct classes. With Type I we have a binary system, of which one member (A) is more massive than its companion (B); this means that B grows in mass, while A declines to become a very small, dense white dwarf, made up chiefly of the element carbon. The situation is then reversed, as B evolves, becomes a red giant in its turn, and loses material back to A; it reminds one of a cosmical tug-of-war. When the white dwarf has accumulated so much material that it has become unstable, the carbon detonates, and in only a few seconds the luckless white dwarf is completely destroyed. At its peak, the explosion may send out a hundred thousand million times as much energy as the Sun, which is as great as that of all the stars in a normal galaxy put together.

A Type II supernova is quite different, and is due to the collapse of a very massive star which has used up all its nuclear 'fuel' and has produced an iron-rich core. When energy production suddenly stops, the outer layers collapse on to the core. Up to this moment the core has been made up of atoms consisting of protons, which are positively charged, and electrons, whose charge is negative; when the core collapses the protons and electrons are squeezed together to form neutrons, which have no electrical charge (the positive protons cancel out the negative electrons). The resulting density is so great that over two and a half thousand million tons of neutron star material could be crammed into a matchbox, and the temperature soars to around 100,000 million°C. There follows a shock-wave which disrupts the star, blowing most of its material away into space to leave only the neutron core.

A neutron star is bizarre by any standards. According to current theory the outer layer is solid iron in crystalline form, below which comes neutron-rich 'fluid'. The deep interior may again be solid, and is assumed to be made up of 'hyperons', even more fundamental than neutrons, but about which our ignorance is more or less complete. It has been said that a neutron star is like a stale raw egg, with a thin outer shell and various peculiar fluids inside. It is highly unlikely that anyone will go there to find out; if you could stand on the surface of a neutron star, you would have a hundred thousand million times your Earth weight.

A neutron star is probably no more than about a dozen miles in diameter. It must cool down after its formation, and within a thousand years the temperature will be no more than a million degrees C; eventually all its energy is used up, and nothing is left but an inert lump.

Of the four supernovæ seen in our Galaxy in the past thousand years, one — the star of 1054 — has left the remnant known as the Crab Nebula, which contains a pulsar. The other three were of Type I; we can still detect weak radio waves coming from the positions in which they were seen.

Star Legends and Sky Gods

This is alleged to be a picture of Ptolemy (Claudius Ptolemaeus), the last great astronomer of Classical times, who died around AD180. Whether or not it is a true likeness must be regarded as dubious!

It has been said that the starlit sky is a vast picture book. In a way this is true enough, because the stars do seem to be arranged in definite patterns; almost everyone can recognise the seven stars of the Great Bear or Plough, the characteristic shape of Orion, or — from the southern hemisphere — the Centaur and the Southern Cross. Legends about them have sprung up, and have been told and re-told throughout the ages. Whenever I see Orion shining down I am bound to think of the celestial Hunter, with his gleaming Belt and misty Sword, keeping well clear of the menacing Scorpion on the opposite side of the sky.

Our familiar legends are Greek, and the constellations we know are Greek in origin even though we call them by their Latin names. The last great astronomer of ancient times, Ptolemy of Alexandria, left us a list of forty-eight constellations, all of which we still recognise even though their outlines have been adjusted and new groups have been added (in particular Ptolemy, who spent his whole life in Egypt, was never able to see the groups round the south celestial pole). But the constellation patterns have no real significance, because the stars are at very different distances from us, and we are dealing with nothing more significant than line of sight effects. In Orion, for example, Rigel lies at almost twice the distance of Betelgeux, and there is absolutely no real connection between the two.

If we had followed a different system, our star maps would also look very different even though the stars themselves would be exactly the same. Even today there have been some 'improvements' which seem illogical. Consider the constellation Pegasus, commemorating the mythological flying horse which carried the hero Bellerophon aloft in his pursuit of a particularly nasty fire-breathing dragon known as the Chimæra. Pegasus, which dominates the evening sky during autumn in the northern hemisphere of the Earth, is marked by four moderately bright stars (Alpheratz, Scheat, Markab and Algenib) which make up a square pattern. For some reason or other the International Astronomical Union, the controlling body of world astronomy, has moved Alpheratz from Pegasus into

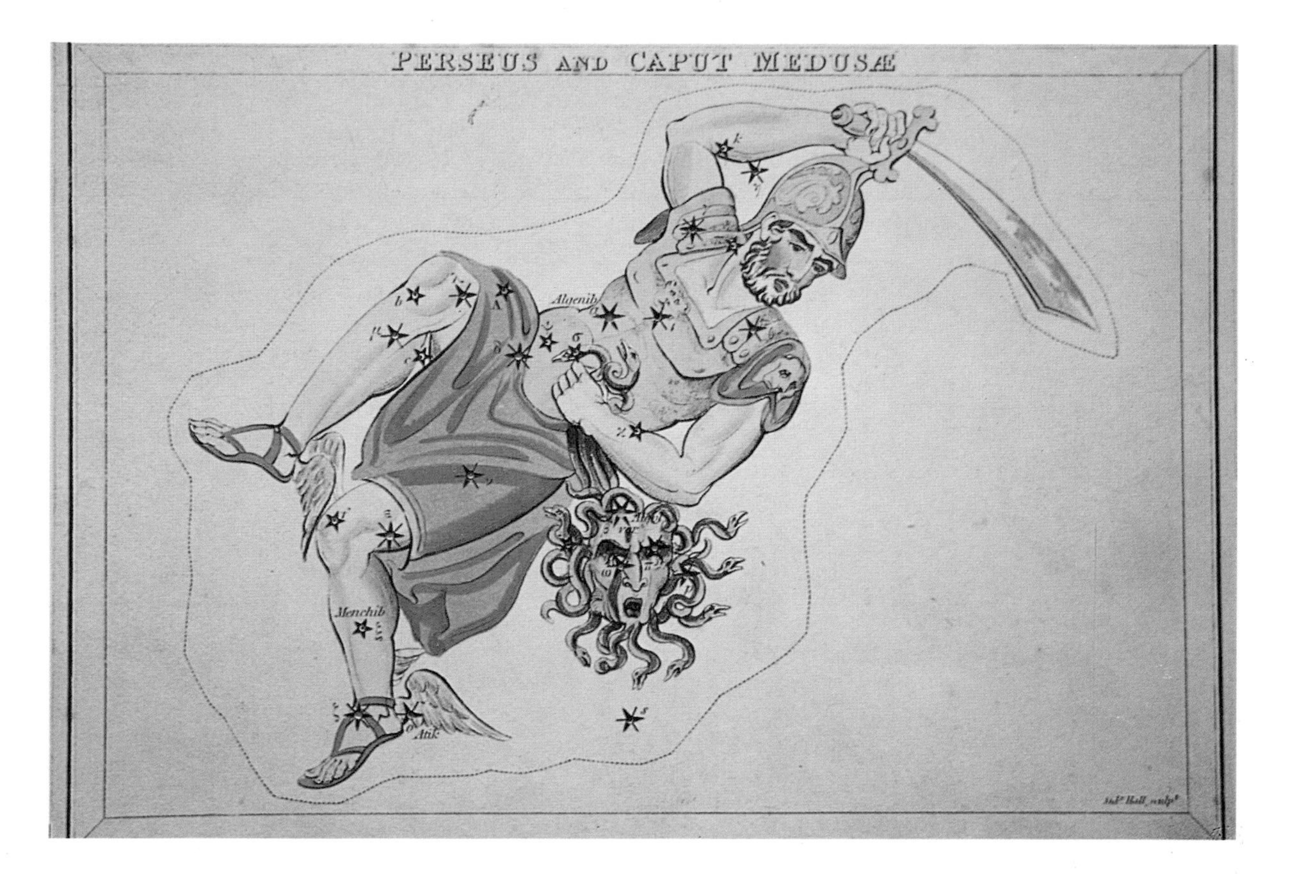

the neighbouring constellation of Andromeda. Similarly, Al Nath in the well-marked constellation of Auriga (the Charioteer) has been given a free transfer into the adjacent Taurus (the Bull). Not that it really matters, but at least it underlines the fact that our constellation names are entirely arbitrary. Had we adopted the Egyptian system, for instance, we would have had no Great Bear and no Bull, but we would have had a Cat and a Hippopotamus.

To us, the mythological stories are attractive legends; to the ancients they were much more than that, and it was natural to associate them with the stars. Much the most famous legend concerns Perseus and Andromeda. It is said that the sea-god Neptune sent a monster to ravage the land of King Cepheus, whose wife, Queen Cassiopeia, had been unwise enough to boast that her daughter Andromeda was more beautiful than the sea-nymphs. As the nymphs were Neptune's daughters, this was clearly tactless. (In Greek mythology, of course, the sea-god was Poseidon, but we always use the Roman name.) Consulting the Oracle, Cepheus learned that the only way to appease the god was to chain his daughter to a rock by the sea-shore to await the arrival of the monster. This was duly done, but fortunately the hapless princess was seen by the intrepid hero Perseus, who was on his way home after disposing of yet another monster, Medusa the Gorgon, who had a woman's head and snaky hair, and whose glance could turn any living thing to stone. Perseus, mounted on winged sandals loaned to him by Mercury, swooped down, turned the monster to stone by showing it the Gorgon's head,

Above:
Figure of Perseus from an old star map

Below:
Figure of Andromeda from an old star map

Astronomical murder

David Fabricius, a minister of the Dutch Reformed Church, was an expert amateur astronomer. In 1617 he preached a sermon in which he hinted darkly that he knew the identity of a member of his congregation who had stolen one of his geese. Evidently he was right, because he was murdered before he could divulge the name of the culprit.

His son Johann was also a good observer, and an independent discoverer of sunspots, but he died in 1616, a year before his father, at the early age of twenty-nine.

and then, in the best tradition, married Andromeda and lived happily ever after. This is one of the few mythological legends with a pleasant ending!

All the characters are to be found in the sky. Cepheus and Cassiopeia are not far from the north celestial pole, with Cassiopeia, marked by five brightish stars arranged in a W or M pattern, considerably more conspicuous than her husband. Perseus is nearby, with Andromeda marked by a line of stars leading off from the Square of Pegasus. Cetus, the sea-monster, is in the southern sky, though it is true that most catalogues relegate it to the status of a harmless whale.

There is an interesting aside here. The head of Medusa, the Gorgon, is marked by the star Algol, which normally shines of the second magnitude — about as bright as the Pole Star. Every two and a half days Algol gives a long, slow 'wink', taking four hours to fade and remaining dim for a mere twenty minutes before taking another four hours to recover. It is known as the Demon Star, and there is a tradition that the old sky-watchers knew about its peculiar behaviour. In fact, it seems that they did not; the variations in light were not reported until 1669, when attention was drawn to them by an otherwise obscure Italian astronomer named Montanari. All the same, it is a strange coincidence that Algol was associated with

MIRA, 'THE WONDERFUL'

In 1595 a Dutch amateur astronomer, David Fabricius, was making some observations when he recorded a moderately bright star in the constellation of Cetus, the Whale. A few weeks later he could no longer see it, but, surprisingly, he never followed the matter up.

When Johann Bayer was compiling his catalogue, in 1603, the star was back, and Bayer gave it the Greek letter Omicron. Again it vanished, and again nobody took much notice. It was only in 1638 that another Dutch observer, Phocylides Holwarda, found that it appeared and disappeared regularly, so that it was visible without optical aid for only a few weeks in every year. It was a genuinely variable star, and it was named Mira, 'The Wonderful'.

At its best Mira can become brighter than the Pole Star, and there are reports that it has even been known to equal Aldebaran in the Bull, though generally it remains much fainter (according to my observations there were bright maxima in 1987 and 1989, but not in 1988). In a famous book published over a century ago the French astronomer Amédée Guillemin claimed that at minimum it vanished so completely that it could not be found even with a telescope, but this is wrong; it can always be followed with a fairly small telescope, or even with powerful binoculars.

Mira is a red giant star, well advanced in its life-story, and has become unstable, so that it is changing its output of energy as it swells and shrinks. The period between one maximum and the next is about 331 days, but this is not absolutely constant, and may vary by several days either way. It is about 100 light-years away, and is much larger and more luminous than the Sun. It has a faint companion which is classed as a sub-dwarf, hot and bluish-white and much denser than the Sun and apparently intermediate in type between a Main Sequence star and a white dwarf.

Mira was the first variable star of its type to be discovered, but by now many thousands are known, though only a few attain naked-eye visibility. Among them is Chi Cygni, in the Swan, which has a much longer period (407 days) and is very red indeed. All the Mira variables are pulsating, and there are also red stars which have smaller variations and less regular periods, among them Betelgeux in Orion.

There are so many variable stars in the sky that professional astronomers have no time to keep watch on them all, and amateur observers do valiant work in providing extra information. In fact, variable star study has now become one of the most important branches of present-day amateur astronomy.

demonic powers in past years. It is not genuinely variable; it is the brightest member of a class of stars known as eclipsing binaries, and the 'winks' are caused when the fainter member of the pair passes in front of the brighter, partially blocking out its light.

Another legend concerns Hercules, the prototype hero, who was sentenced to perform twelve labours, each of which would have defeated most demigods. Among his victims were an extremely unpleasant monster known as Hydra, which had a hundred heads, and also an outsize cat, the Nemæan Lion. Hydra is actually the largest constellation in the sky, though it is also one of the dullest and has only one reasonably bright star. Leo (the Lion), on the other hand, is one of the most prominent of the constellations of the Zodiac, and as seen from Britain dominates the evening sky all through spring. Hercules himself is large, but decidedly dim and formless. One feels that in view of his reputation, he deserved something rather better.

Orion, the Hunter, is predictably prominent. Among his retinue are his two Dogs, Canis Major and Canis Minor, while below his feet is a Hare (Lepus). According to legend, Orion boasted that he could kill any creature on earth, but met his end when a scorpion crawled out of the ground and stung him in the heel. When Orion was restored to life and placed in the heavens, it seemed only fair to put the Scorpion (Scorpius) there too — but on the far side of the sky, so that the two could never meet. From Britain, Orion and Scorpius can never be above the horizon at the same time.

Note, by the way, that though the old constellation names are of Greek origin, some other names have a decidedly modern flavour. We have Telescopium (the Telescope), Microscopium (the Microscope), Octans (the Octant), Antlia (the Air-pump) and others. These groups were added long after Ptolemy's time, mainly between 1600 and 1700. Other constellations, with barbarous names such as Psalterium Georgianum (George's Lute), Globus Ærostaticus (the Balloon), Sceptrum Brandenburgicum (the Sceptre of Brandenburg) and Machina Electrica (the Electrical Machine) were mercifully deleted when the International Astronomical Union undertook its major revision of the sky-map. Most of the deletions were fully justified, though I rather regret the demise of Quadrans (the Quadrant) and Felis (the Cat). Quadrans has not been completely forgotten, because the meteors seen coming from that part of the sky in early January are still known as the Quadrantids.

Names of individual stars are mainly Arabic, given about a thousand years ago. Some of them are tongue-twisters. For example, 'Betelgeux' may be spelled in several different ways ('Betelgeuse' and 'Betelgeuze' are other variants) and there are also different pronunciations; I remember that Sir James James, the great astrophysicist who was also one of the first BBC broadcasters on astronomy, insisted on calling it 'Beetlejuice'. The real meaning seems to be 'the Armpit of the Central One', while Rigel marks the Hunter's foot.

Nowadays most of the individual names have fallen into disuse, except for the very brightest stars plus a few special cases such as

A cardinal rule

On the night of 23 February 1987 Ian Shelton, a Canadian astronomer working at the Las Campanas Observatory in Chile, was taking long-exposure pictures of the Large Cloud of Magellan, which is the nearest of the really important outer galaxies, and is 170,000 light-years away. (European astronomers always bemoan the fact that it lies so far south in the sky and is invisible to them.) When he developed his plates, he found to his surprise that there was a bright spot near the gas-patch known as the Tarantula Nebula. Thinking that it might be a flaw in the plate, he went outside the dome, looked up at the Cloud, and there was the bright spot: a supernova, the first to become visible with the naked eye since 1604. It was seen at about the same time by an amateur at the Observatory, Oscar Duhalde.

Ironically, Robert McNaught at the Siding Spring Observatory in Australia had photographed the Cloud on the previous night, but had not developed it. When he did so, there was the supernova. He had forgotten one of the cardinal rules of astronomy: never put off examining your observations!

Opposite top:
Figures of Hydra, Corvus, Crater and Noctua from an old star map. Hydra was the great serpent killed by Hercules; it and Corvus (the Crow) and Crater (the Cup) are still to be found on modern maps, but Noctua (the Night Owl) has disappeared

Opposite below:
Figure of Gemini from an old star map

Mizar in the Great Bear, Algol in Perseus, and Mira, the variable star in Cetus (the Whale) which can become quite prominent, though on average it is visible with the naked eye for only a few weeks in every year. I sometimes think that this is rather a pity, because the names are romantic, but it is true that they tend to be rather cumbersome; what can one say of, for instance, Zubenelchemale, Alkalurops, Sadalbari and Azelfafage?

Mythology is a fascinating study, and I have always been particularly interested in it, mainly because it is so closely associated with the sky. But it has also been suggested that some of the ancient legends are founded upon something which is not pure fancy, and have even influenced human thought. Among these is the tale of Oannes, half man and half fish, who is said to have emerged from the Persian Gulf and taught the early Mesopotamians how to read and write. Our main account comes from a celebrated historian whose name is given as Berossus. He lived in the time of Alexander the Great, and is said to have written a detailed history of Mesopotamia, though since the original has not come down to us, and we have to rely upon second-hand versions, it is impossible to tell how reliable it is.

I mention Oannes here because of the suggested link with a tremendous outburst in the sky which undoubtedly took place around the year 8,000BC. It was a supernova, and it burst forth in the southern constellation of Vela, the sails of the ship Argo. Vela is part of one of Ptolemy's original forty-eight constellations, Argo Navis, the ship which carried Jason and his companions in their rather unprincipled quest of the Golden Fleece. Argo was so large that during the modern revision of the constellations, it was chopped up into a keel (Carina), a poop (Puppis) and sails (Vela). Most of it, including the brightest of its stars — Canopus, in Carina — is too far south to be seen from Britain. A few stars of Puppis rise above our horizon, but of Vela we can see nothing.

In Vela there is a misty patch which is known as the Gum Nebula, not because it is sticky but because it was carefully studied by an Australian astronomer named Colin Gum (who, sadly, died young; he was killed in a skiing accident). The Gum Nebula contains a pulsar, which sends out radio waves and can also be seen as a faint optical object only just above the visible threshold of our largest telescopes. The presence of a pulsar proves that the Gum Nebula is is a supernova remnant, and we can judge that the outburst happened about 10,000 years ago. This means that there were no true astronomers to observe it, but it must have been so brilliant that it would have caused general excitement and, no doubt, alarm. From Mesopotamia it would have been low down, and its light, reflected upon the ocean surface, must have been striking. If the star were regarded as divine, there could well have been the impression of a god walking on water, and this, according to a researcher named George Michanowsky, accounts for the Oannes tale.

The only corroborative evidence comes from a cuneiform tablet, known to scholars as BM-86378, which is 2in wide x 3in high. It

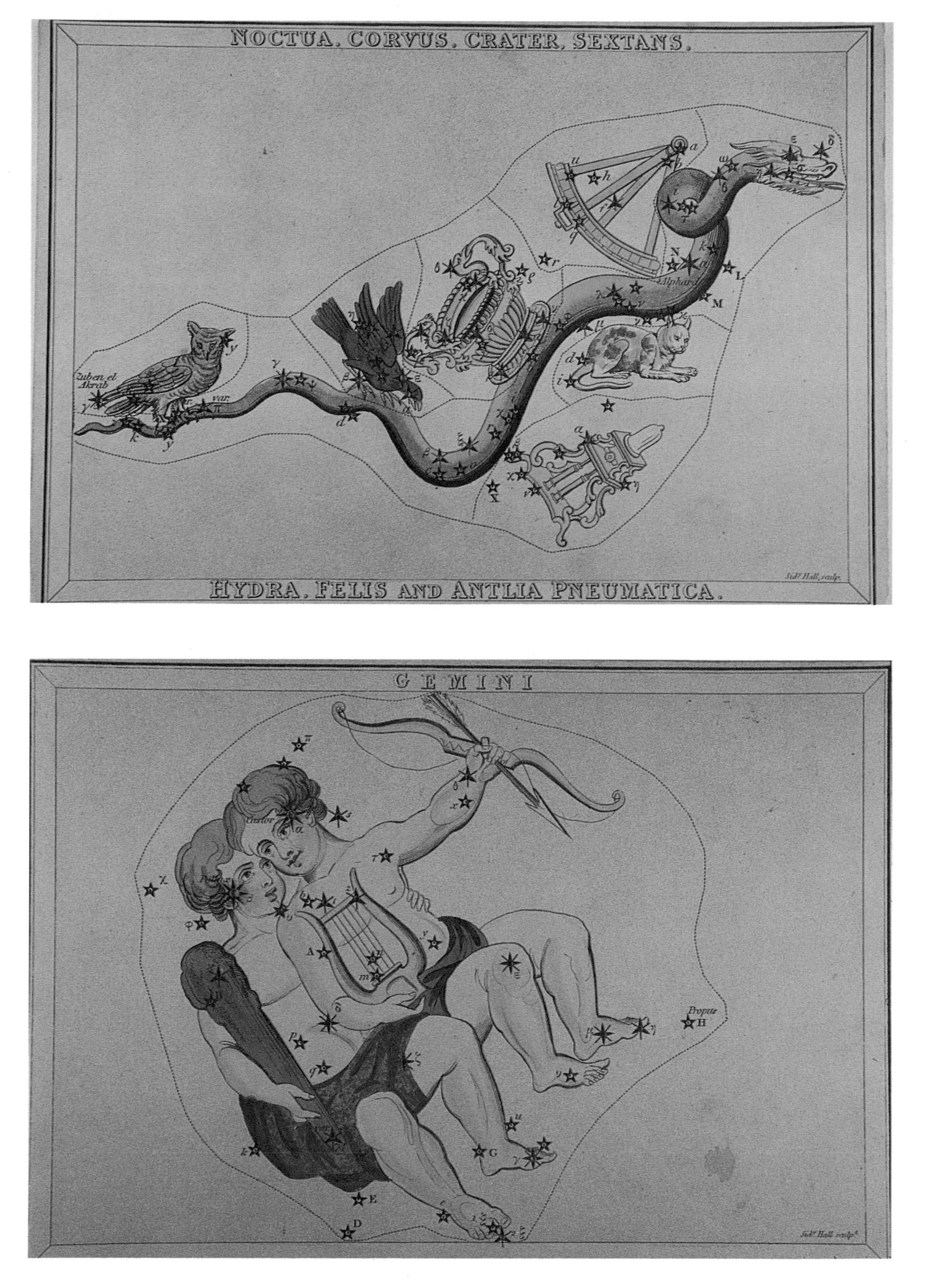
NOCTUA, CORVUS, CRATER, SEXTANS.
HYDRA, FELIS AND ANTLIA PNEUMATICA.
Alphard
Zuben el Akrab
GEMINI
Castor
Pollux
Propus

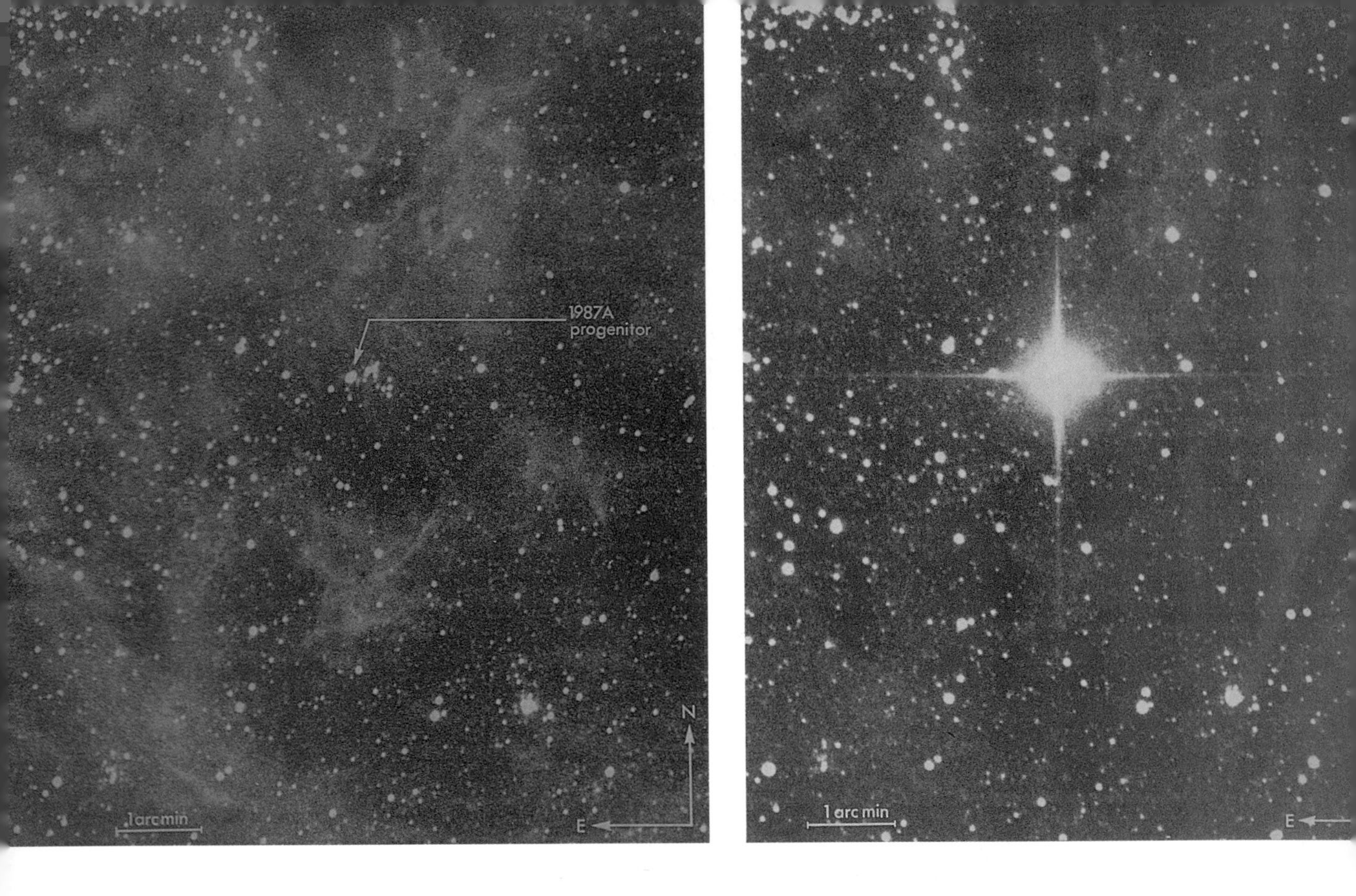

Above:
Supernova 1987A: Progenitor, taken before the outburst. The progenitor star (a blue supergiant) is arrowed

Above right: Supernova 1987A: Maximum, the same region, taken after the outburst. The transformation is truly remarkable (Las Campanas)

seems to be a copy of a Mesopotamian star catalogue, and is written partly in Sumerian (which nobody has yet succeeded in deciphering) and partly in Akkadian or Assyro-Babylonian. The suggestion is that the catalogue contains an entry relating to something overpoweringly brilliant in a position corresponding to that of the present-day Gum Nebula. It is all very uncertain, but we cannot rule out a link with the Vela supernova and the legend of Oannes.

Certainly the outburst must have been magnificent. Supernovæ are the most colossal explosions known in Nature, and at peak brightness may be at least a hundred thousand million times as luminous as the Sun. The brightest in historical times was seen in 1006, in the constellation of Lupus (the Wolf), which may have shone as brightly as the quarter-moon, though unfortunately it is poorly documented. Since then we have had supernovæ in our Galaxy in 1054, 1572 and 1604, all of which became bright enough to be seen with the naked eye in broad daylight; the 1054 star has left the remnant we know as the Crab Nebula, which is of immense importance to astronomers because it sends out radiations in almost all parts of the electromagnetic spectrum. But if our calculations are correct, the Vela supernova of 8000BC was very much brighter than any of these, and for a while it would have turned night into day.

The last galactic supernova appeared before the invention of the telescope, and modern astronomers would dearly like the chance to study one at reasonably close quarters, so what are the prospects? The best 'supernova candidate' we know is again in the far south. Eta Carinæ, in the keel of the old Ship, is very massive and apparently unstable. For a while in Victorian times it outshone every star in the sky apart from Sirius, and though it is at present just below naked-eye visibility it seems to be partly shrouded by interstellar dust, so that it may brighten up again at any moment. It is around 6,000 light-years away, and is several millions of times more powerful than the Sun. If it does 'go supernova' it will indeed be glorious, and this is bound to happen eventually, though perhaps not for a very long time. I well remember my first view of it through a telescope; it looks like an orange blob, quite unlike a normal star, and it is associated with nebulosity. Recent studies indicate that it may be made up of several components, very close together. Eta Carinæ is unique in our experience, and astronomers keep a watchful eye upon it. If it turns into a supernova, it will certainly rival the Vela outburst of so long ago, but presumably it will be well documented, so that we can hardly

Erratic Eta Carinæ

Eta Carinæ is the most erratic of all variable stars. Our first definite record of it comes from Halley, who saw it during his sojourn in St Helena in 1677 and rated it as of the fourth magnitude. It was rated magnitude 2 by Lacaille in 1751, by Brisbane in 1825 and by Fallows in 1830; by 1827 it had brightened to magnitude 1, and on 16 December 1837, Sir John Herschel, at the Cape, made it equal to Alpha Centauri, the third brightest star in the whole of the sky. Following a slight decline there was a new flare-up in 1843, when Eta Carinæ actually outshone even Canopus, so that it was inferior only to Sirius. After 1844 there was a steady fading, and by 1867 the magnitude had dropped to 6. Since then it has hovered on the fringe of naked-eye visibility. Quite apart from its intrinsic variability, we have to reckon with the changing amount of dust and gas between the star and ourselves, so that a new outburst may occur at any moment.

Region of Eta Carinæ. Eta Carinæ, a possible supernova candidate, lies in a rich area; unfortunately it is too far south to be seen from Britain

'a little star'

In the early part of August 1885 a Hungarian lady, the Baroness de Podmaniczky, was holding a house-party at her home. One of her guests was a professional astronomer named de Kövesligethy. There was a 3½in telescope available, and to interest her guests the Baroness took it out on to the lawn that evening and turned it toward the Andromeda Spiral. Suddenly she said that she could see 'a little star', and de Kövesligethy and the other guests agreed, but they believed it to be due to the moonlight, and took no action; the telescope was put away, the house-party went on, and the 'star' was forgotten. Yet what they had seen was a supernova, over two million light-years away. We now call it S Andromedæ.

Actually, the Baroness and the doctor were not the first to see it. On 17 August it had been picked up from Rouen by Dr Ludovic Gully, but was taken to be a telescopic fault. On 20 August it was seen by Dr Hartwig, who was on the staff of the Dorpat Observatory in Estonia, but the Observatory

continued opposite

have another Oannes legend. Meanwhile, can we link any other old tales to astronomical phenomena about which we can be reasonably confident?

Of course the classic case is that of the Star of Bethlehem, which has caused endless argument now for almost two thousand years. The trouble is that we know so little about it. It is mentioned only once in the Bible, in the Gospel according to St Matthew, Chapter 2. Matthew tells us that the Wise Men from the East came to Herod, saying 'Where is he that has been born King of the Jews? for we have seen his star in the east, and are come to worship him.' Verses 7-10 run as follows:

> Then Herod, when he had privily called the wise men, inquired of them diligently what time the star appeared. And he sent them to Bethlehem, and said, Go and search diligently for the young child; and when ye have found him, bring me word again, that I may come and worship him also.
>
> When they had heard the king, they departed; and lo, the star, which they saw in the east, went before them, till it came and stood over where the young child was. When they saw the star, they rejoiced with exceeding great joy.

That is all that St Matthew tells us; the other Gospels do not mention the Star at all, and to make matters worse we are by no means certain about our dates. At least, we know that Christ was not born on 25 December, 1AD, and a better guess is 4BC. Incidentally, 25 December was not celebrated as Christmas Day until the fourth century, and by that time the real date had long since been forgotten, so that our Christmas is wrong too.

If the Star of Bethlehem had been an ordinary object, such as a planet, everyone would have seen it, and it would have caused no particular interest; if the Wise Men could have been deceived by Jupiter or Venus, they would not have been very wise. Therefore, if there is any truth in the story, the Star must have been unusual. A brilliant supernova or even a nova would have been observed by contemporary astronomers, but the records tell us nothing of the kind; neither were there any bright comets (Halley's Comet came back at least seven years too early). One favourite explanation is that of a conjunction between two planets, lying apparently side by side in the sky, but this again would have lasted for some time, quite apart from the fact that there would have been no exceptional eastward motion. All in all, we have to admit that there is no plausible scientific explanation for the Star of Bethlehem, and it is most unlikely that we will ever find out much more.

Coming back to the Bible, we find that the Book of Ezekiel begins with a remarkable account of what has been claimed to be an alien spaceship carrying four astronauts, each of whom had four faces and four wings. Fiery wheels were also in evidence. It must have been a most upsetting spectacle, and one can hardly be surprised that Ezekiel fell upon his face; in similar circumstances most people would probably have done the same.

The space-craft theory has been championed recently by Erich von Däniken, who goes so far as to maintain that God himself is the result of a genetic experiment carried out by visitors from another world, who landed here (possibly to re-fuel), fertilised some members of the primitive Earth species, and went away, later returning to see the results of their work. Von Däniken does concede that the fertilisation theory is 'full of holes' (no comment!) and perhaps we had better leave it at that.

By the time we come to pyramidology, Atlantis, Easter Island and flying saucers, we have well and truly left the realm of science, but I must backtrack and say something about the Cosmic Serpent, because it has been discussed by two very eminent modern astronomers, Drs Victor Clube and Bill Napier.

Clube and Napier go back to ancient legends, but with a much more rational approach than those of the von Däniken school. They concentrate upon comets, which, of course, have always been regarded as unlucky, and they cite the Roman historian Pliny, who, in his *Natural History,* wrote: 'A terrible comet was seen by the people of Ethiopia and Egypt, to which Typhon, the king of that period, gave his name; it had a fiery appearance and was twisted like a coil and it was very grim to behold; it was not really a star so much as what might be called a ball of fire.' Clube and Napier believe that various disasters of near-historic times can be put down to impacts of meteorites or comets, and they postulate the former existence of a vast comet, the Cosmic Serpent, which caused mayhem at regular intervals until it finally broke up. Its remnants can still be seen today as asteroids, one of which, Hephaistos, could be the surviving main part of the Serpent even though it has long since lost all its dust and gas, so that nothing remains except its inert core.

This means that the ancient dragon myths have a basis in reality, and the Biblical Flood could have been due to an encounter with a swarm of cometary bodies around the year 2500BC — in which case some of our dating of early historical events will be in need of drastic revision. One paragraph in the book is particularly significant: 'If short-period comets were indeed sky-gods, and the comet which we are now calling the Cosmic Serpent came spectacularly close to the Earth at intervals, then the desirability of predicting the returns would be clear: astronomy would grow out of theology.'

I admit to being unconvinced, but it is certainly true that the Earth must be struck by massive bodies now and then. We cannot discount the theory that a major impact around 65,000,000 years ago caused such a change in the Earth's climate that the dinosaurs, which had ruled the world for so long, were unable to cope with the new conditions and quickly died out.

Whether any of this is true or not, we can still enjoy the legends we see written for us in the sky. They never pall, and when I look up at Perseus, Cassiopeia, Cepheus and Andromeda there are times when it almost seems possible to see the beautiful princess chained to the rock.

continued

director had not seen it for himself, and refused to let Hartwig announce it in case there had been a mistake. Another early observer was Isaac Ward of Belfast. Then, on 25 August, it was seen by Max Wolf from Heidelberg, and its reality was established. At its best it almost reached naked-eye visibility, but nobody realised its exceptional nature, and it was generally taken to be an ordinary nova in the same line of sight as the Andromeda Spiral. G. F. Chambers, observing with a 6in refractor on 3 September, commented that in his view 'the star had nothing to do with the nebula'.

The star soon faded, but in 1988 R.A. Fesen and A.J.S. Hamilton, using the 158in reflector at Kitt Peak in Arizona, detected its remnant as a tiny darkish patch cutting out the light of stars beyond. So we have not lost S Andromedæ, but astronomers would have been happier if it had appeared in 1985 rather than in 1885.

THE CONSTELLATIONS

NAME		BRIGHTEST STAR(S)	NOTES
Andromeda	Andromeda		
Antlia*	The Air-pump		
Apus*	The Bird of Paradise		
Aquarius	The Water-bearer		Zodiacal
Aquila	The Eagle	Altair	
Ara*	The Altar		
Aries	The Ram	Hamal	Zodiacal
Auriga	The Charioteer	Capella	
Boötes	The Herdsman	Arcturus	
Cælum*	The Graving Tool		
Camelopardalis**	The Giraffe		
Cancer	The Crab		Zodiacal
Canes Venatici**	The Hunting Dogs	Cor Caroli	
Canis Major	The Great Dog	Sirius	
Canis Minor	The Little Dog	Procyon	
Capricornus	The Sea-goat		Zodiacal
Carina*	The Keel	Canopus	
Cassiopeia**	Cassiopeia		
Centaurus*	The Centaur	Alpha Centauri, Agena	
Cepheus**	Cepheus		
Cetus	The Whale		
Chamæleon*	The Chameleon		
Circinus*	The Compasses		
Columba*	The Dove		Partly invisible from Britain
Coma Berenices**	Berenice's Hair		
Corona Australis*	The Southern Crown		
Corona Borealis	The Northern Crown	Alphekka	
Corvus	The Crow		
Crater	The Cup		
Crux Australis*	The Southern Cross	Acrux, Beta Deneb, Crucis	
Cygnus	The Swan		Partly circumpolar from Britain
Delphinus	The Dolphin		
Dorado*	The Swordfish		
Draco**	The Dragon		
Equuleus	The Foal		
Eridanus	The River	Achernar	Partly visible from Britain
Fornax*	The Furnace		
Gemini	The Twins	Pollux, Castor	Zodiacal
Grus*	The Crane		
Hercules	Hercules		
Horologium*	The Clock		
Hydra	The Watersnake	Alphard	
Hydrus*	The Little Snake		
Indus*	The Indian		

THE CONSTELLATIONS

NAME		BRIGHTEST STAR(S)	NOTES
Lacerta**	The Lizard		
Leo	The Lion	Regulus	Zodiacal
Leo Minor	The Little Lion		
Lepus	The Hare		
Libra	The Balance		Zodiacal
Lupus*	The Wolf		
Lynx**	The Lynx		
Lyra**	The Lyre	Vega	
Mensa*	The Table		
Microscopium*	The Microscope		
Monoceros	The Unicorn		
Musca	The Fly		
Norma*	The Rule		
Octans*	The Octant		South Polar
Ophiuchus	The Serpent-bearer	Rasalhague	Crosses Zodiac
Orion	Orion	Rigel, Betelgeux	
Pavo*	The Peacock		
Pegasus	The Flying Horse		
Perseus	Perseus	Mirphak, Algol	
Phœnix	The Phœnix		
Pictor*	The Painter		
Pisces	The Fishes		Zodiacal
Piscis Austrinus	The Southern Fish	Fomalhaut	
Puppis*	The Poop		
Pyxis*	The Compass		
Reticulum*	The Net		
Sagitta	The Arrow		
Sagittarius	The Archer		Zodiacal
Scorpius	The Scorpion	Antares	Zodiacal. Partly invisible from Britain
Sculptor	The Sculptor		
Scutum	The Shield		
Serpens	The Serpent		Divided into two parts
Taurus	The Bull	Aldebaran	Zodiacal
Telescopium*	The Telescope		
Triangulum	The Triangle		
Triangulum Australe	The Southern Triangle		
Tucana*	The Toucan		
Ursa Major**	The Great Bear		'The Plough'
Ursa Minor**	The Little Bear	Polaris	North Polar
Vela*	The Sails		
Virgo	The Virgin	Spica	Zodiacal
Volans*	The Flying Fish		
Vulpecula	The Fox		

** denotes that the constellation is more or less invisible from Britain*
*** that it is more or less circumpolar from Britain*

The Sky from Pole to Pole

Opposite:
16in telescope at the Science Centre, Singapore. This is on an English mounting, but because Singapore is almost on the equator the axis is virtually horizontal, giving it a most unfamiliar look!

Mizar, the famous double star in the Plough; it is officially known as Zeta Ursæ Majoris. Here, Mizar itself is shown as double with the more distant Alcor close by. 20in lens (f/6.3), exposure 30sec (Commander H. R. Hatfield)

During 1989 I paid my first visit to Singapore. At the Science Centre there a new observatory has just been completed, with a 16in reflector as its main telescope, plus a splendid exhibition in the Centre itself. Of course, Singapore is a large city, and there is too much light pollution to please an astronomer – it is a world-wide problem – and at present not much can be done about it. At least Singapore has the advantage of being within a degree of the equator, so that the whole of the sky can be seen during the course of the year. For example, it is possible to see the Great Bear on one side of the sky and the Southern Cross on the other.

We in Britain miss the far southern stars, just as New Zealanders can never see our Great Bear, but one way to become familiar with 'inaccessible' stars is to go to a planetarium. Here, an artificial sky is projected on to the inside of a large dome, and the effect is remarkably realistic. All sorts of phenomena can be produced, including such rarities as total eclipses of the Sun, brilliant comets and spectacular auroræ, while it is also a simple matter to 'speed up' the movements of the planets and show how they behave. However, nothing can be an adequate substitute for the glory of the real sky, so let us go on another journey in imagination, starting at the north pole and working our way through to the Antarctic.

THE VIEW FROM THE POLE

Most people know that the Earth's equator divides the world into two hemispheres. Lines of latitude are reckoned from the equator (latitude 0°) and extend to the north pole (+90°) and the south pole (-90°). In the same way, the celestial equator divides the sky into two hemispheres, and we have the equivalent of latitude, though we call it by a different name: declination. This means that the north celestial pole has a declination of +90°, while the declination of the celestial equator is zero.

The north pole of the sky is closely marked by the bright star Polaris, in Ursa Minor or the Little Bear; the actual declination is +89°16', so that it is not exactly at the pole. If we start our journey at the north pole of the Earth, keeping a watchful eye out for large white

RIGHT ASCENSION AND DECLINATION

The position of any point on the Earth's surface is defined by its latitude and longitude; for example, the position of my observatory at Selsey, in Sussex, is latitude 50°43'49"·25, longitude 00°47'41"·25W. In the sky, the equivalent co-ordinates are declination and right ascension. Declination is simply the angular distance of the object north or south of the celestial equator, measured on what we call the celestial sphere — an imaginary sphere surrounding the Earth and with the same centre. Right ascension, however, is slightly more complicated.

Longitudes on Earth are measured from the Greenwich meridian, which was accepted as the zero by international agreement in 1884. We need a zero for the celestial equivalent of longitude, and the point we use is the Vernal Equinox, or First Point of Aries.

During its annual journey round the sky, the Sun traces out a path against the stars which we call the ecliptic. It is inclined to the equator by 23½°, and so each year the Sun crosses the equator twice: the points where the ecliptic and equator cut are termed the equinoxes, and the spring or vernal equinox is the point where the Sun crosses the equator in late March, moving from south to north. It is called the First Point of Aries because originally it lay in the constellation of Aries, the Ram. (The effect known as precession has now shifted it into the adjacent constellation of Pisces, the Fishes, but we still use the old name.) The right ascension of a star is its angular distance from the First Point of Aries, measured eastward.

It might be thought that the value should be given in degrees of arc, but actually it is more convenient to use hours, minutes and seconds of time. Because the Earth makes one rotation per day (approximately twenty-four hours) the First Point of Aries reaches its highest point, and 'culminates', once a day. The right ascension of a star is the difference in time between the culmination of the First Point of Aries, and the culmination of the star. Thus Betelgeux in Orion culminates 5hr 55min 10sec after the First Point of Aries has done so, and the right ascension of Betelgeux must be 5hr 55min 10sec.

Apart from the slight shifts due to precession, the right ascensions and declinations of the stars remain constant, but those of the Sun, Moon and planets are changing all the time.

bears, we will find that Polaris is directly overhead, and will remain there.

Note, in particular, that the altitude of Polaris is equal to our latitude as seen from the north pole: 90°. In fact, the altitude of Polaris is always equal to your own latitude from the point of observation; from London (latitude approximately 51°) Polaris willl be 51° above the horizon.

Now for the celestial equator. Its declination is zero; from the north pole, its altitude will also be zero — in fact, the celestial equator will run all round the horizon. The stars will seem to move round Polaris in circles parallel with the horizon, and no star will either rise or set. The northern hemisphere of the sky will be seen all the time; the southern hemisphere, never. This means that a constellation such as Orion, which is crossed by the celestial equator, will be cut in half by the horizon.

When the Sun is moving north of the celestial equator, as it does between late March and late September, it will remain above the horizon as seen from the north pole, which explains the six-months' 'day'. For the rest of the year there is no direct sunlight at all, though for much of the time the sky glows with brilliant displays of auroræ.

Sooty stars

We know of stars which sometimes hide themselves behind veils of soot. One of these is R Coronæ, in the small but easily found constellation of the Northern Crown, not far from Arcturus. Normally it is just below naked-eye visibility, so that binoculars will show it easily, but at unpredictable intervals it fades, becoming so faint that it cannot be seen at all without the help of a powerful telescope. What apparently happens is that soot builds up in the star's atmosphere, and blocks part of the light; eventually the radiation from below blows the soot away, and R Coronæ returns to its usual brightness. R Coronæ is not unique, but certainly these 'sooty stars' are very rare.

GOING SOUTH

Now let us move away from the north pole, leaving the Esquimaux and bears behind, and see what happens to the sky. As our latitude decreases, Polaris sinks downward. Further south than around latitude 66° north we can never see the 'midnight sun'; we have to stay in places such as Alaska or North Norway. Lerwick, in the Shetland Isles, is at latitude 60°, and this is not sufficiently far north for the midnight sun, though in June it is quite easy to go outdoors at midnight and read a newspaper.

THE VIEW FROM BRITAIN

Over mainland Britain we have real darkness for part of the night, even in June, and much of the sky is available at one time or another. The latitude of my own home, Selsey in Sussex, is 50° north, so that in theory I can see any star which is further north than declination –40°. Similarly, any star which is north of declination +40° will be 'circumpolar', and will never set, since even at its lowest — below the Pole Star — it will still remain above the horizon.

From Britain — and also from all other places of around the same latitude, such as parts of the United States — the main circumpolar constellations are the two Bears, Ursa Major and Ursa Minor, together with Cassiopeia. Ursa Major is often nicknamed the Plough, or, in America, the Big Dipper; its seven main stars make a pattern which cannot be mistaken. Mizar, the second star in the 'tail', has a much fainter star, Alcor, close beside it. Curiously, the Arabs of a thousand years ago regarded Alcor as a test of keen eyesight, but there is nothing difficult about it today. If you have a telescope, you will see that Mizar itself is made up of two stars, and there is another, dimmer star between the main pair and Alcor.

Ludwig's star

The faint star between Mizar and Alcor is of the eighth magnitude. It can be seen with powerful binoculars, and almost any telescope will show it, but it is well below naked-eye visibility. It was first reported in 1723, and courtiers of the Emperor Ludwig V named it 'Sidus Ludovicianum' because they imagined that it had appeared specially for their ruler's benefit.

There have been suggestions that it is variable, and that a thousand years ago it was much brighter than it is now, so that when the Arabs spoke of Alcor being 'a test for keen vision' they did not really mean Alcor at all, but Ludwig's Star. This seems unlikely; all the same , it is not easy to explain why the Arabs found Alcor so elusive, when today it is a very easy object to anyone with even moderately good sight.

The Double Cluster in Perseus: The Sword-Handle. Two open clusters side by side. The clusters are visible with the naked eye as a misty patch, and binoculars show them well. 10in reflector (Bernard Abrams)

The two 'end' stars of the pattern, Dubhe and Merak, are known as the Pointers, because they show the way to Polaris. Look closely, and you see that while Merak is white, Dubhe is somewhat orange, showing that it has a lower surface temperature and is further advanced in its life-story.

The Little Bear is easy to identify; it curves down over the Great Bear's tail, but it is much fainter, so that strong moonlight will drown most of it. On the far side of Polaris is the well-marked W of Cassiopeia, which is crossed by the Milky Way. Cassiopeia and the Great Bear are on opposite sides of Polaris, and at about the same distance from it — so that when Cassiopeia is high up, the Great Bear is low down, and vice versa.

On a winter's evening Ursa Major will be seen 'standing on its tail' in the north-east, with Cassiopeia high in the north-west. Looking south, you will see Orion, which is probably the most spectacular constellation in the entire sky. It contains two exceptionally brilliant stars, the orange-red Betelgeux and the glittering white Rigel; in the centre of the pattern are the three stars of the Hunter's Belt, and below the Belt is the Sword, which looks like a misty patch. Use binoculars, and you will have an excellent view of the Great Nebula itself, which is well over a thousand light-years away. There are many of these 'stellar nurseries' in the sky, but the nebula in Orion's Sword is the most famous of them.

Orion is very convenient in showing the way to other stars and star-groups. Downward, the Belt points to Sirius, in Canis Major (the Great Dog), which is outstandingly bright — though you will need

Præsepe, the famous 'Beehive' cluster in Cancer, taken with a 10in reflector. This cluster covers a wide area, and is probably best appreciated with binoculars (John Fletcher)

a powerful telescope to glimpse its white dwarf companion. Upward, the Belt points to Aldebaran in Taurus (the Bull), which is orange-red, like Betelgeux; extending from Aldebaran is a little V-formation of stars, making up the open cluster of the Hyades. Follow the line still further, and curve it somewhat; you will come to the most beautiful star-cluster in the sky, the Pleiades or Seven Sisters. At first glance the cluster looks like a patch of haze, but if you look more closely you will make out individual stars in it. Normal-sighted people can see at least seven, and if you can manage a dozen you are doing very well indeed. Binoculars bring out many more, and altogether the cluster contains several hundred members.

Almost overhead lies Capella, in Auriga (the Charioteer). It is unmistakable because of its brilliance, and because there is a small triangle of stars close beside it. Capella is of the same colour and temperature as the Sun, but whereas the Sun is a yellow dwarf Capella ranks as a yellow giant. Auriga itself has a sort of quadrilateral form, and adjoins Perseus, whose most interesting star is the 'Demon' Algol.

Also in Orion's retinue are the Twins, Castor and Pollux, in the constellation Gemini. Pollux, the brighter of the two, is decidedly orange, while Castor is white; in fact Castor is a multiple system, made up of four main components and a dwarf pair. Gemini itself is marked by lines of stars extending from Castor and Pollux in the general direction of Betelgeux. Like Auriga, it is crossed by the Milky Way, and is very rich. Between the Twins and Sirius you can see yet another bright star, Procyon in Canis Minor (the Little Dog). Winter evenings also provide us with the Square of Pegasus, setting in the west, and Leo, the Lion, coming into view in the east.

Opposite:
Messier 31, the Andromeda Galaxy. 200in photograph (Palomar Observatory)

Orion, the Hunter, crossed by the celestial equator; of its two leaders, Betelgeux is orange-red, Rigel pure white (Hatfield Polytechnic Observatory)

Spring evenings bring a marked change. Orion is fast disappearing into the twilight, though Sirius remains on view for some time after sunset; the Great Bear is high, and Leo has taken over the dominant position in the south, marked by a curved line of stars of which Regulus is the brightest member; this is the so-called Sickle, while the rest of Leo consists of a well-marked triangle further to the east. There is a minor mystery associated with the brightest star of the triangle, Denebola. In ancient times it was said to be as bright as Regulus, but it is now obviously fainter, so that either the star has faded or else (more probably) the old records are wrong.

Another star which comes into prominence during early spring is Arcturus, in Boötes (the Herdsman), which you can find by following round the curve of the Great Bear's tail. Arcturus is very brilliant, and is of a lovely light orange colour. It was once thought to be one of the most luminous stars known; this is not so, but it is still the equal of 115 Suns put together. Its distance is 36 light-years, so that we in 1991 see it as it used to be in 1955. Not far from it is the little semicirclet of stars marking Corona Borealis, the Northern Crown. Few constellations bear the slightest resemblance to the objects after which they are named, but Corona is one of the exceptions. Another bright star on view is Spica, in Virgo (the Virgin), which lies further along the curve traced from the Bear through Arcturus. Virgo itself is Y-shaped, though apart from Spica its stars are not particularly bright.

Between Castor and Pollux on one side and the Sickle of Leo on the other lies the dim Zodiacal constellation of Cancer, the Crab. It is not too easy to identify, but it does contain Præsepe, one of the brightest of all open star-clusters. You can see it easily with the naked eye on a dark night. It is often nicknamed the Manger or the Beehive; I have never understood why the ancient Chinese referred to it as 'the Exhalation of Piled-up Corpses'!

By summer evenings we have lost Orion, together with his retinue. Ursa Major is high in the north-west, and Leo is setting. Near the overhead point is a lovely blue star, Vega in Lyra (the Lyre or Harp), which takes up the position occupied during winter evenings by Capella. If you see a brilliant star straight overhead, it can only be one of these two.

Vega makes up a huge triangle with two other bright stars, Deneb in Cygnus (the Swan) and Altair in Aquila (the Eagle). Cygnus is often nicknamed the Northern Cross, for obvious reasons. One member of the X-pattern, Albireo, is fainter than the rest and further away from the centre, so that it spoils the symmetry, but to make up for this Albireo is a glorious double star, with a golden yellow primary and a bluish companion. Almost any telescope will show the two stars separately, though with the naked eye Albireo looks quite undistinguished.

I never tire of looking at Albireo; to my mind it is the most superb of all the coloured doubles. The two components are genuinely associated, but they are a long way apart, and are highly luminous, since even the blue star is at least 120 times as powerful as the Sun.

Tycho Brahe's oversight

The Andromeda Spiral was known in ancient times; in 905 the Arab astronomer Al-Sufi called it 'the Little Cloud'. The first telescopic observation seems to have been made by Simon Marius in 1611, who described the soft glow as looking like 'the light of a candle seen through horn'. Yet very curiously, Tycho Brahe, the best observer of pre-telescopic times, overlooked it completely.

If there are any planets in the Albireo system, their skies must be truly magnificent.

Very low over the southern horizon during summer evenings lie two of the constellations of the Zodiac, Scorpius (the Scorpion) and Sagittarius (the Archer), both of which are crossed by some of the richest parts of the Milky Way. Scorpius is led by the red supergiant Antares, the 'Rival of Mars', so named because of its fiery hue —

STAR MAGNITUDES

The stars are divided into grades or magnitudes according to their apparent brightness. The scale works in the same way as a golfer's handicap, with the more brilliant performers having the lower values; thus magnitude 1 is brighter than magnitude 2, 2 is brighter than 3, and so on. Stars of magnitude 6 are the faintest which can normally be seen with the naked eye on a clear night. Note that a star's apparent magnitude has little or nothing to do with its real luminosity, because the stars are at very different distances from us; a star may look conspicuous either because it is very near, because it is very powerful, or a combination of both factors.

With the naked eye it is possible to detect differences of a tenth of magnitude. For example, look at three stars in Orion's Belt. Alnilam, the middle star, is of magnitude 1·7, while Alnitak, the southern or lower member of the Belt, is 1·8; you can see that Alnilam is slightly but perceptibly the brighter of the two.

Stars above magnitude 1 have zero or even negative magnitudes; thus Sirius is –1·5. On this scale Venus can reach –4·4, while the Sun is –27. At the other end of the scale, good binoculars will reach down to magnitude 8 or 9, while our largest telescopes can record objects fainter than magnitude 28. Stars of above magnitude 1·4 are usually said to be of 'the first magnitude'. They are:

Star	Constellation	Magnitude	Real luminosity, Sun=1	Colour
Sirius	Canis Major	–1.5	26	White
Canopus*	Carina	–0.7	200,000	White
Alpha Centauri*	Centaurus	–0.3	1.5	Yellowish
Arcturus	Boötes	–0.1	115	Orange
Vega	Lyra	0.0	52	Bluish
Capella	Auriga	0.1	70	Yellow
Rigel	Orion	0.1	60,000	White
Procyon	Canis Minor	0.4	11	Yellowish
Achernar*	Eridanus	0.5	780	Bluish-white
Betelgeux	Orion	var	15,000	Orange-red
Agena*	Centaurus	0.6	10,500	Bluish-white
Altair	Aquila	0.8	10	White
Acrux*	Crux Australis	0.8	3200+2000	Bluish-white
Aldebaran	Taurus	0.8	100	Orange
Antares	Scorpius	1.0	7500	Red
Spica	Virgo	1.0	2100	White
Pollux	Gemini	1.1	60	Orange
Fomalhaut	Piscis Australis	1.2	13	White
Deneb	Cygnus	1.2	70,000	White
Beta Crucis*	Crux Australis	1.2	8200	White
Regulus	Leo	1.3	130	White

** indicates that the star is invisible from Britain. Next in order come Adara in Canis Major (1.5) and Castor in Gemini (1.6). The magnitude of the Pole Star is 2.0; that of Sigma Octantis, the southern pole star, is only 5.5.*

Sigma Octantis

The south pole star, Sigma Octantis, is very faint, and is not of much use to navigators. However, it is not so insignificant as it looks; it is 120 light-years away, and is six hundred times as powerful as the Sun.

Ares, the Greek war-god, was the equivalent of the Roman Mars. Unfortunately we Britons never see the Scorpion well, because it is always low down, and part of it never rises at all. Following it round is the Archer, with the glorious star-clouds which mask our view of the mysterious region at the centre of the Galaxy. Sagittarius has no particular shape; I have never been able to work out why so many people liken it to a teapot!

During evenings in autumn the triangle formed by Vega, Deneb and Altair is still on view, with the Great Bear at its lowest in the north and the W of Cassiopeia almost overhead. The main autumn constellation is Pegasus, and well below it, not far above the horizon, is Fomalhaut in the Southern Fish, which is much brighter than you might think; from North Scotland it barely rises.

Leading off from the Square of Pegasus is the line of stars marking Andromeda. Look carefully above the second of the two brightish stars in the line, and you should be able to make out the misty blur of the Great Spiral, which is the nearest of the really large galaxies. It is over 2,000,000 light-years away, and is the most remote object which can be definitely seen without optical aid. The question is often asked: 'How far can one see without a telescope?', the answer being 'Well over twelve million million million miles', because this is the distance of the Andromeda Spiral.

FURTHER SOUTH

Now let us recommence our journey, moving southward toward the equator. The Pole Star sinks lower and lower; by the time we reach Athens — or, in America, San Francisco — Polaris is a mere 38° above the horizon, and during summer we can see the whole of the Scorpion and the Archer. From the Canary Islands we can also see Canopus in the keel of the Ship, but to see the famous Southern Cross we must go further south still; Hawaii will do quite well. By now, of course, our familiar groups such as the Great Bear and Cassiopeia are no longer circumpolar, but spend part of each twenty-four-hour period below the horizon.

THE EQUATOR

Now the whole of the sky is available at some time during the year; the celestial equator passes straight overhead, and the celestial poles are on opposite horizons.

During January evenings, Orion is at the zenith or overhead point; Capella is high in the north, and the whole of the Hunter's retinue is on view, but Ursa Major is almost invisible, as only part of it remains above the northern horizon. In the south we have the Ship, led by Canopus; Sirius is high, while low in the south-west it is possible to glimpse Achernar in Eridanus, the River.

April evenings are of special interest. Orion has gone, but in the east Scorpius is rising; the overhead position is occupied by Virgo, followed by Leo. In the north the Great Bear is at its best, lying at an angle which to Britons seems unfamiliar; the 'tail' points upward to Arcturus. The southern aspect is dominated by the Southern Cross

The summer triangle

Vega, Deneb and Altair make up a large triangle. In a television Sky at Night programme many years ago (in 1958) I referred to this as 'the Summer Triangle', because from Britain it is at its best during summer evenings. To my surprise everyone started to use the term, which has now become an accepted part of astronomical language even though it is quite unofficial and the stars are in different constellation (Vega in Lyra, Deneb in Cygnus and Altair in Aquila). Moreover, southern countries such as Australia see the Triangle at its best during evening in winter!

and the Centaur, with Sirius now very low in the south-west and Canopus grazing the horizon.

By July, the Southern Cross is again low during the evening, and there is little to be seen of Ursa Major in the north, but the large triangle made up of Vega, Deneb and Altair is very prominent, with Vega almost due north. Arcturus remains visible in the north-west. But it is the Scorpion which now rules; it is high in the south, and in its way it is almost as magnificent as Orion. This is also the best time to see the star-clouds of Sagittarius, which are far more splendid than they can ever be as seen from the latitude of Britain.

October evenings bring us the Square of Pegasus, not far from the zenith; the Vega-Deneb-Altair triangle is setting in the west, while Cassiopeia is well on view in the north. Orion is starting to come into view, with Achernar and Fomalhaut high. This is the time of the 'Southern Birds', of which the easiest to find is Grus, the celestial Crane; neither can you mistake the two Clouds of Magellan, which are independent galaxies, and which are very prominent naked-eye objects. Northern astronomers always regret that the Clouds lie so far south in the sky.

THE SKY FROM THE SOUTH

It is time to move on once more. From Darwin, in the Northern Territory of Australia, we can still see the Great Bear for part of the year, though it is never very high. By the time we reach the latitude of Sydney, which is much the same as that of Cape Town in South Africa or Montevideo in South America, we have almost lost it, but the Southern Cross is circumpolar and Canopus almost so, while the south celestial pole has risen to 34° above the horizon.

Unfortunately there is no Antipodean equivalent of Polaris, and the south polar area is depressingly blank. The nearest naked-eye star to the pole is an obscure one known only as Sigma Octantis. It is none too easy to locate even when you know just where to look for it, and any mist or moonlight will drown it. This is one way in which northerners have the advantage; southern navigators would willingly exchange their Sigma Octantis for our Polaris!

In summer evenings (that is to say, around January) Orion is high, but this time it is Rigel which is in the upper left of the pattern, with Betelgeux to the lower right, while the Belt points upward to Sirius and downward to Aldebaran. The usual retinue can be seen, but Capella is always low, and is not so very far above the northern horizon.

Canopus, near the overhead point, is a superb star. It does not look so brilliant as Sirius, but it is far more powerful; Sirius is as luminous as twenty-six Suns, but according to one reliable estimate it would take 200,000 Suns to match Canopus. Its surface is not much hotter than that of the Sun, and it should be slightly yellowish, but to me it always appears pure white. Not far from it lies the strange, erratic variable Eta Carinæ, our potential supernova candidate, together with open clusters, gaseous nebulæ and superbly rich star-fields.

The Southern Cross is now well in view, gaining altitude in the

Tachard at the Cape

The story of southern-hemisphere astronomy really began in 1685, when the French Jesuit priest, Father Guy Tachard, visited the Cape of Good Hope when he was on his way to Siam, and set up a temporary observatory. He was well received by the Dutch, under Governor Simon van der Stel, and from the outset he did his best to be courteous. Indeed, he may have been over-anxious, and at the outset there was a truly Gilbertian muddle with regard to the firing of the correct number of gun salutes, as Tachard related in his book *Voyage de Siam des Pères Jesuites*:

'It was agreed upon that the Fort should render Gun for Gun when our Ship saluted it. This Article was ill explained, or ill understood, by these Gentlemen, for about ten of the Clock my Lord Ambassador having ordered seven Guns to be fired, the Admiral answered with only five Guns, and the Fort fired none at all. Immediately the Ambassador sent ashore again, and it was determined that the Admirals Salute should pass for nothing, and so the Fort fired seven Guns, the Admiral seven Guns, and the other Ships five, to salute the Kings Ship, which returned them their Salutes, for which the Fort and Ships gave their Thanks.'

This little contretemps having been overcome, the travellers went ashore to a warm greeting. During his stay Tachard re-measured the longitude of the Cape, studied the movements of the satellites of Jupiter and also made some stellar observations, during which he discovered the double nature of the brightest star in the Southern Cross. When the time came to say good-bye, the Dutch presented the astronomers with 'Presents of Tea and Canary Wine', while Tachard left his hosts a microscope and a burning-glass. He never returned to the Cape, but he must have carried away very happy memories of it and its people.

south-east. It is more like a kite than a cross; there is no central star to the X, as with Cygnus, but to make up for this it is very packed, with three brilliant stars and one which is moderately bright. The leader, Acrux or Alpha Crucis, is a fine double, separable with any telescope; adjoining the main pattern is the Coal Sack, an apparently starless area which is nothing more nor less than a dark mass of dust and gas, hiding everything which may lie beyond it. The Cross also contains a lovely cluster, the so-called Jewel Box, which contains one red star shining out against its bluish companions.

Following the Cross are the Pointers, Alpha and Beta Centauri.

CLUSTERS OF STARS

Spread throughout the Galaxy we find definite clusters of stars whose members make up genuinely associated systems, and which unquestionably have a common origin. The clusters are of two main types: open (or 'loose') and globular.

Of the open clusters, much the most famous is that known as the Pleiades, or Seven Sisters, in Taurus (the Bull). It has been known since very early times; in Homer's *Odyssey* (Book 5) it is said that Odysseus 'sat at the helm and never slept, keeping his eyes upon the Pleiads', and the cluster is even referred to in the Bible. Naturally, there is a mythological legend about it. It is said that the Pleiades were seven beautiful sisters who were taking a casual stroll in the forest when they were espied by the hunter Orion, who pursued them with intentions which were anything but honourable. To save them from a fate worse than death they were snatched up by the gods and placed in the sky, though it must be admitted that they are still rather too near Orion to feel entirely comfortable!

How many of the Pleiades can you see with the naked eye? I once carried out an experiment, enlisting the help of viewers of my television programme *The Sky at Night,* and found that the average number was indeed seven, though some people could see more. The record is reputedly held by a nineteenth-century German astronomer named Eduard Heis, who could count nineteen. The brightest member is Alcyone (Eta Tauri), followed by Electra, Atlas, Merope, Maia and Taygete, with Celæno, Pleione and Asterope on the limit of naked-eye visibility; it must be added that Pleione is somewhat variable, and occasionally throws off 'shells' of material, so that it is decidedly unstable. The leading stars of the Pleiades are hot, bluish-white and very young by cosmical standards; there is a great deal of nebulosity in the cluster, so that fresh stars are still being formed. The distance is just over 400 light-years.

The Hyades, also in Taurus, are very different, and are overpowered by the brilliant orange light of Aldebaran — which is unfortunate, because Aldebaran is not a genuine member of the cluster, and simply happens to lie in the same line of sight, about half-way between the Hyades and ourselves. The cluster is much more scattered than the Pleiades, and is considerably closer to us.

Several other open clusters are visible without optical aid, notably Præsepe in Cancer, the lovely Jewel Box in the Southern Cross, and the double Sword-Handle in Perseus (not to be confused with the Sword of Orion). But open clusters are not tightly bound by gravitation, and eventually they will be disrupted, so that their member stars will drift away and the clusters will lose their identities.

The globular clusters are quite different. Here we have tightly bound, symmetrical systems which may contain more than a million suns. They lie around the main part of the Galaxy, and make up a sort of 'outer framework'; all are very remote, lying at distances of the order of 20,000 light-years or more. About 140 are known in the Galaxy, but only three are visible with the naked eye, Omega Centauri and 47 Tucanæ in the far south and the great northern cluster in Hercules. Their leading stars are red giants which have evolved off the Main Sequence, and so the globulars themselves seem to be very old indeed. When viewed telescopically their outer parts can be resolved easily enough, but near the centre the stars are much more closely packed, though even they are still so widely separated in space that direct collisions or even 'close encounters' must be rare.

To an observer living on a planet moving round a star near the middle of a globular cluster, the sky would be magnificent; the stars would be only light-weeks apart rather than light-years, and many of them would be brilliant enough to cast shadows, so that there would be no true darkness at all. Against this, an astronomer on such a world would find it very difficult to learn much about the universe beyond his own particular cluster!

Omega Centauri, the finest of all globular clusters, too far south to be seen from Britain. It is a prominent naked-eye object

Alpha Centauri (which, strangely, has no official proper name) is the nearest of all bright stars beyond the Sun; the second Pointer, Agena, is remote and powerful, so that the two are quite unconnected. The rest of the Centaur is rich, and almost surrounds the Southern Cross. Look for Omega Centauri, which is not a star, but a globular cluster — a huge symmetrical system of stars, more than a million in all, packed so closely near the centre that it is impossible to see them individually.

Higher up, and to the south-west, is the brilliant Achernar, nicknamed 'the Last in the River'. The south celestial pole lies about midway between Achernar and the Cross, and this is also the region of the two Clouds of Magellan, which we know to be independent galaxies even though they give the misleading impression of being broken-off parts of the Milky Way. The Large Cloud lies at a distance of 170,000 light-years, and is so bright that it can be seen even during the time of full moon; it contains objects of all types, ranging from giant and dwarf stars to clusters, nebulæ, novæ and even supernovæ. The Small Cloud is not so bright, but it too is easy to see without a telescope, and almost in front of it is another globular cluster, 47 Tucanæ. It is easy to think that the cluster and the Cloud are connected, but this is not so; the Cloud lies far in the background.

Autumn evenings (that is to say, around April) give us Leo in the

Opposite:
The planetarium sky, taken inside the dome of the Armagh Planetarium

north, Virgo near the zenith, and Arcturus in the north-east. The Southern Cross and the Centaur are now very high, with Sirius and Canopus dropping in the south-west and Scorpius reaching a respectable altitude in the south-east. By winter evenings (July) it is the Scorpion which is dominant, almost overhead, while the star-clouds of Sagittarius are superb. The Cross and the Centaur are still

THE DEVELOPMENT OF THE PLANETARIUM

The ancestor of the planetarium is the orrery, which is a moving model; the Sun is in the middle, and the planets can be made to revolve round it at the correct relative speeds. Hand-operated orreries were made in the seventeenth century by Ole Rømer, the Danish astronomer who was the first man to measure the velocity of light. Rather later, a London clockmaker, John Rowley, made a clockwork orrery for Charles Boyle, Earl of Orrery — hence the name. There were even some large orreries set up in the roofs of buildings, one of which was constructed between 1774 and 1781 by a Dutchman, Eise Eisanga, at his house at Franeker in Holland. It was weight-driven, and is still in working order.

Next came the Gottorp Globe, made in Denmark about 1660 by Andreas Busch, a mechanic employed by Duke Frederik III of Holstein, together with Oleander, the Court Mathematician. They realised that to be given an impression of the sky, the audience would have to look upward. Their hollow globe was 11ft in diameter, weighing almost 3½ tons, and it could be rotated by means of water power, making one complete turn in twenty-four hours — though the operator could always speed it up for demonstration purposes. Inside the globe, hung from its fixed axis, was a round table, together with a bench large enough to seat ten people. The star patterns and constellation figures were shown on the inside of the sphere, and the globe was lit, from within, by two oil lamps. When the globe was rotated, therefore, the audience could see the stars and constellation figures drifting across the 'sky' in a very realistic manner. Admittedly the Gottorp Globe used no projector, and was not a planetarium in the modern sense of the term, but it proved to be very popular.

Other globes followed, notably one made in 1911 by W.W. Atwood for a museum in Chicago. Here the diameter was 15ft, and the globe was made of very thin sheet-iron, so that it weighed only about 500lb. Holes drilled in the metal made it possible to show about seven hundred stars when the globe was illuminated from outside. As with earlier models, the brighter stars were represented by larger holes, so that the effect was remarkably striking. The globe was rotated by electrical power, and for the first time planets were shown; since the planets move around they had to be represented by several holes in different positions, so that the holes not in use could be blocked out. There were also differently shaped luminous disks to represent the crescent, half, three-quarter and full moon.

Soon after the end of World War I Professor Walther Bauersfeld, of the Zeiss optical works in Germany, introduced an entirely new principle. He did away with the rotating dome, and showed the stars on the inside of a stationary dome by means of a special projector. Bauersfeld's device was remarkably complicated; every star had to be exactly positioned, and there had to be two 'ends' to the projector — one for the northern hemisphere of the sky, the other for the southern — making the instrument look rather like a dumbbell. The Sun, Moon and planets had their own projectors, so that they could be moved around correctly, and all sorts of extra refinements were introduced.

When Bauersfeld's first planetarium was opened, at Jena (East Germany) in 1925, it proved to be a tremendous success, and others soon followed. At first they were all made by Zeiss, but then the Japanese joined in, as well as the Spitz company in America, while a few large planetaria were 'home-made'. Today they are widespread, and most large cities have them. A few are highly specialised, such as the Northern Lights Planetarium at Tromsø in Norway, which presents particularly good displays of auroræ.

The London Planetarium at Madame Tussaud's, in Marylebone Road, was opened in 1958. I was invited to become the first Director, but I declined, though in 1965 I did become the first Director of the new planetarium at Armagh in Northern Ireland, and stayed there for three years until it was in full working order.

Other British planetaria are being planned. If you have not been to one of these 'Theatres of the Stars', I strongly recommend a visit.

with us, but Canopus is missing. Look north, and you will see Vega, Altair and Deneb, but Deneb at least grazes the horizon. Finally we come to spring evenings, in October, when it is the turn of the Cross to dip down to the horizon, while Canopus rises in the south-east and the Scorpion sets in the south-west. Pegasus is on view in the north; near the zenith is Fomalhaut, which shows up to advantage, and which surprises Britons by its brightness.

THE SOUTH POLE

Further south still, from New Zealand, we lose the last trace of our familiar northern groups. From Invercargill even Capella is out of view, though Canopus has become circumpolar. We are nearing the end of our trip.

From the Antarctic, Sigma Octantis is overhead, and once again the celestial equator girdles the horizon, with Orion cut in half — only this time it is Rigel which can always be seen against a darkened sky, Betelgeux never. I have never been to Antarctica, but am told that the clarity of the atmosphere is almost unrivalled, and of course we also have the Southern Lights, which flash and flicker throughout the six-months' period of darkness. At the moment, plans are being made to set up a permanent observatory there. It would certainly be worth visiting.

How many of these varied skies have you seen? There must be many people who have travelled to far-away lands and have never bothered to look upward. If you decide to go, then do not forget to take your star-maps and your binoculars. I can assure you that you will find plenty to see.

Ancestor of a planetarium

In 1669 a new sort of globe was constructed by Erhard Weigel, of the University of Jena in Germany. In its way it was not too unlike the Gottorp Globe, but it was made of iron sheeting, and had some unusual features. The stars were not fixed to the inside, as at Gottorp, but were produced by small holes in the iron sheeting, so that when the outside lights were turned on — plunging the globe's interior into darkness — the audience sitting inside could see the stars as pinholes in the dome, the brightness of the artificial star depending upon the size of the pinhole.

A model Earth was also set up inside the globe, containing working models of the volcanoes Etna and Vesuvius which gave off steam, flames, and what were described as 'pleasant odours'. Hail, rain, wind, thunder and lightning could also be reproduced. If all these special effects were turned on at once, the spectators must have emerged in a somewhat bemused state!

Who Can Hear Us?

One of my hobbies is music. I am very much of an amateur, and I have never had a music lesson in my life, but this does not stop me from enjoying it. Quite recently, when I was playing a Strauss waltz on the piano, an interesting thought struck me. How would it be if my piano had no effective notes except those of the middle octave?

Certainly I would not have been able to produce a waltz, or anything else, and neither would an expert pianist do better. We need all the notes, from the low to the high. Until little more than half a century ago, the astronomer was faced with the same sort of problem; he could use only the 'middle octave' of the total range of wavelengths, and therefore he was hopelessly handicapped.

The full electromagnetic spectrum extends from the ultra-short gamma-rays through X-rays, ultra-violet, visible light, infra-red, microwaves and then to radio waves. All are vibrations of the same basic nature, which is not easy to appreciate when we can actually see only a few of them — those of our middle octave.

What we normally call 'white' light is not really white at all, but is made up of all the colours of the rainbow. In other words it is a mixture of wavelengths, the wavelength being the distance between one wave-crest and the next. Ripples in water have wavelengths of several inches at least, but the wavelengths of light are measured in the tiniest fractions of a millimetre. Red light has the longest wavelength and violet the shortest; in between come orange, yellow, green and blue. Radiations of wavelength longer than that of red light, or shorter than that of violet, cannot be seen, so that ordinary telescopes cannot help us.

Objects in the sky send out radiations at virtually all wavelengths, but so long as we have to observe from the surface of the Earth we are in difficulty simply because of the screening effects of our air. Virtually no cosmic gamma-rays and X-rays can get through, and until the development of rockets and spacecraft very little could be done; not until 1962 was the first X-ray source identified in the sky. Ultra-violet experimenters are not much better off, and much of our present knowledge is due to one particular satellite, the International Ultra-Violet Explorer, which was launched more than

The Sun and radar

Radar, known originally as radiolocation, was developed during the war as an aid to navigation. In 1942 the research team led by J.S. Hey found that there was some unknown source of interference, and they naturally assumed that the Germans had found out about our radar and were trying to jam it. Actually, the mysterious radio waves came from the Sun!

a decade ago and was still functioning excellently in 1990.

Immediately beyond the long-wave end of the visible band we come to infra-red, which is easy to detect simply by switching on an electric fire; you will feel the infra-red, in the form of heat, well before the bars become hot enough to glow. An infra-red telescope looks very like an optical one, but it has to be used in a very different way. The detecting equipment has to be cooled right down, because so far as these devices are concerned the telescope itself is 'hot', and its signals would swamp the feeble infra-red coming in from space. Again the Earth's air is a nuisance, particularly the water vapour in it. The infra-red astronomer likes to observe from as great a height as he can, which is why he has a fondness for the tops of mountains such as Mauna Kea. But even the great Hawaiian volcano is not lofty enough, and artificial satellites provide the only really satisfactory answer.

One of the most successful satellites ever launched was IRAS, the Infra-Red Astronomical Satellite, which was sent up in 1983 and went on functioning for most of the rest of that year. It detected thousands of new infra-red sources, but perhaps its most significant discovery was that some stars, notably the brilliant blue Vega, are associated with clouds of cool material which is radiating at long wavelengths. The discovery was not anticipated (yet another 'tale of the unexpected!') and in fact the infra-red telescope on board IRAS was simply being calibrated on Vega when the material was found.

According to the co-discoverers, Drs Hartmut ('George') Aumann and Fred Gillett, the material round Vega is in the form of a cloud of particles, many of them much larger than the usual dust particles of

Launch of IRAS, from Cape Canaveral

Right:
Artist's impression of IRAS in orbit (NASA)

Below:
Radio telescope at the Rutherford Appleton Laboratory, Chilton, Oxfordshire, where the IRAS signals were received

ordinary interstellar space. The cloud extends out to over 7,000 million miles from the star, which is around eighty times the distance between the Sun and the Earth, so that there must be a considerable quantity of it. 'If there are small particles round Vega,' Dr Aumann told me, 'there must be large particles also. I was astonished, because it seemed that the total mass was much the same as that of all the planets in our Solar System combined.'

In this case, could Vega have a proper system of planets? There seemed no valid reason why not, though proof was lacking. Similar clouds of cool material were found round other stars, notably Fomalhaut in the Southern Fish, so that apparently they were common. With the southern Beta Pictoris, in the little constellation of the Painter, the cloud has actually been photographed; Drs Bradford Smith and Richard Terrile, using the 100in reflector at the Las Campanas Observatory in Chile, managed to get a good picture of it. They found that there is a disk of material extending to 48,000 million miles from the star, and that the composition seems to be very much like that of the Earth and the other planets in the Sun's family.

Jumping to conclusions is always unwise, and it would be much too bold to claim that 'Beta Pictoris is the centre of a planetary system'. Yet the chances are far from negligible, and in any case it would be both conceited and unintelligent to suggest that our Sun is unique in being attended by a peopled planet.

Obviously, not all stars are suitable. Some of them are variable, so that their luminosity changes quickly, and any circling planet would

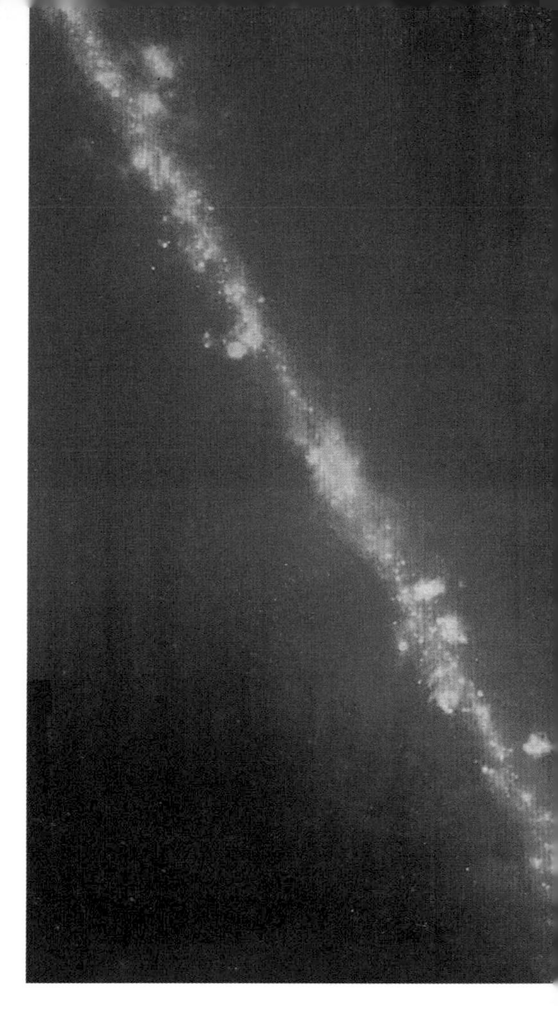

Above: The centre of the Galaxy, from IRAS. This is a false-colour picture (NASA)

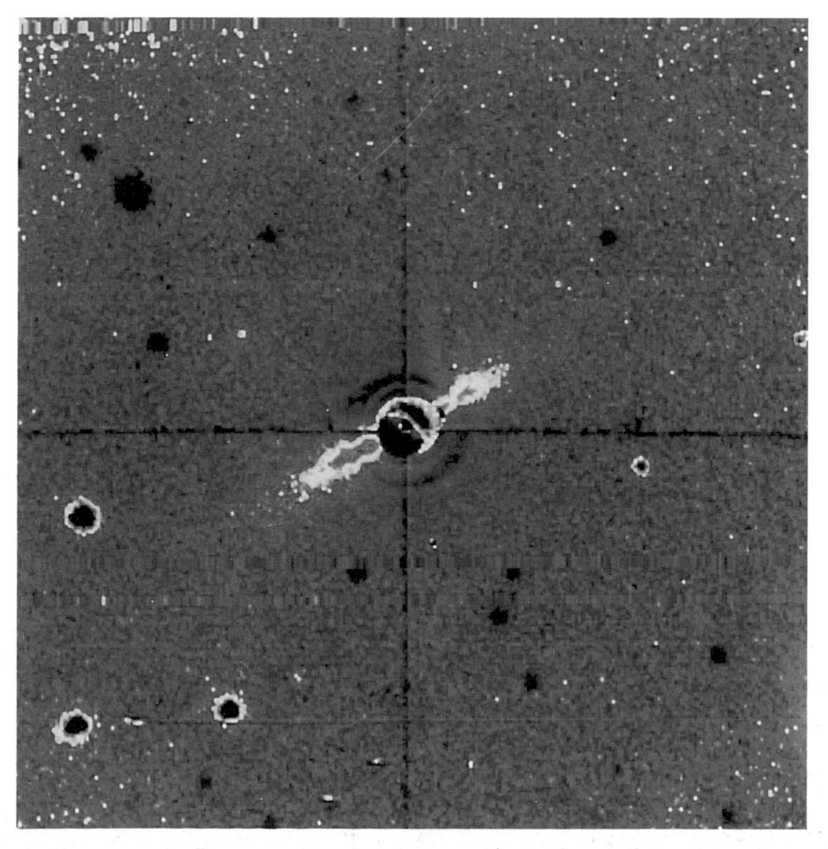

Left: The material associated with Beta Pictoris. The star itself is obscured by the instrumentation (Las Campanas Observatory)

THE OPTICAL AND RADIO WINDOWS

Most of the radiations coming from space are blocked by the Earth's atmosphere. This is fortunate for us, because otherwise no life on Earth could have appeared, but it is undeniably a nuisance to the astronomer, and before the advent of rockets and artificial satellites he was working under very difficult conditions.

Very short wavelengths are now customarily given in nanometres, one nanometre being equal to one-thousand millionth of a metre. The range of wavelengths is roughly as follows (nm = nanometre):

X-rays:	less than 0.01nm
Ultra-violet:	10 – 400nm. (The region between 10 and 120nm is usually called EUV, or Extreme Ultra-Violet)
Visible light:	400 – 700nm
Infra-red:	700nm – 1mm
Microwaves:	1mm – 0.3m
Radio waves:	longer than 0.3m

Most of the short wavelengths are blocked. Between 300 and 750nm we have the 'optical window', where the atmosphere is transparent, but beyond that much of the incoming radiation is again blocked, mainly by water vapour and atmospheric carbon dioxide, though there are a few narrow windows. The main 'radio window', where again the atmosphere is transparent, extends between 20mm to 30m, but still longer wavelengths are reflected back into space by the layer in the Earth's upper air which we call the ionosphere.

PLANETS OF OTHER STARS

Locating planets of other stars is a very difficult matter. Certainly we cannot see them directly, because they are much too faint and much too close to their parent stars (assuming that they really exist, which statistically does seem overwhelmingly probable). We have to use indirect methods which are bound to be rather hit or miss. We can use infra-red methods to detect cool material which is radiating only at long wavelengths, but this merely tells us that 'something' is there, not what it actually is. Finding a bona-fide planet is more of a problem still.

Moreover, there seems to be a sort of transition stage between a planet, which is a non-luminous body sending out no appreciable energy of its own, and a star, which is shining because of nuclear reactions going on inside it. The limiting mass is probably about ten times that of Jupiter, the most massive planet in our Solar System. If the mass is lower than that, the central temperature can never rise high enough to trigger off nuclear reactions.

The earliest results which seemed to be reasonably positive came from studies of a faint red dwarf known officially as Munich 15040, but more generally as Barnard's Star because attention was first drawn to it by the American astronomer Edward Emerson Barnard. Apart from the three members of the Alpha Centauri system it is the nearest of our stellar neighbours, at a mere six light-years, though even so it is much too faint to be seen with the naked eye; its magnitude is 9.5, so that it can just about be glimpsed with good binoculars. It is nicknamed the Runaway Star because of its exceptional proper motion. In only 190 years it moves against its background by a distance equal to the apparent diameter of the full moon — twice as fast as any other known star.

In 1937 Peter van de Kamp, at the Sproule Observatory in America, began making careful measurements of the movements of Barnard's Star. He found that it was 'weaving' its way along, so that presumably it was being pulled out of position by a companion body which was too dim to be seen. The amount of the pull allowed van de Kamp to calculate the companion's mass, and he came to the final conclusion that there were two attendants, both rather less massive than Jupiter — in which case they would definitely be planets, not stars.

It is fair to say that doubts have been cast upon van de Kamp's measurements, but the method itself is perfectly sound, and other stars also have shown signs of being pulled around in the same way. Unfortunately the mass determinations are so uncertain that we cannot tell whether we are dealing with massive planets or with 'failed stars' — that is to say brown dwarfs, with surface temperatures hot enough to make them glow but with cores too cool for them to start shining in the same way as fully-fledged stars. At the moment we have to confess that we have not definitely located either extra-solar planets or brown dwarfs, but no doubt we will find out more in the near future.

have a most uncomfortable climate. Other stars are members of pairs or groups. We can also rule out very powerful stars, which are short-lived by cosmical standards, so that supergiants such as Betelgeux or Antares would long since have gobbled up any planets they may once have had. What we need is a star very like our Sun, shining steadily and not changing much over periods of several thousands of millions of years.

There is no shortage of candidates, inasmuch as solar-type stars are among the commonest in the Galaxy. A good example is Delta Pavonis, in the constellation of the Peacock. It is nineteen light-years away, and is easily visible with the naked eye; in size, temperature and luminosity it is almost exactly a twin of the Sun, so that it might well be expected to have the same sort of system.

The nearest stars which are reasonably similar to the Sun are known as Tau Ceti and Ensilon Eridani. Neither is distinguished enough to merit an individual name; both are rather smaller and cooler than the Sun, and both are roughly eleven light-years from us. There is some evidence that they are attended by low-mass compan-

Thirtieth anniversary

In October 1987 I was privileged to be present at a special ceremony at Jodrell Bank: the thirtieth anniversary of the 250ft 'dish', which was officially re-named the Lovell Telescope. Unknown to Sir Bernard, the legend '57 – 87' had been painted on the underside of the dish, and during the actual ceremony the dish was tilted slowly to reveal the figures; on the previous evening Sir Bernard had had to be lured away from the site in case he noticed what was going on.

The Lovell telescope, 5 July 1987

Paying for Jodrell Bank

The great 250ft 'dish' at Jodrell Bank was finally completed in 1957, after a chequered career. As usual (but in this case, quite excusably) the cost of the telescope had been badly underestimated, and the whole venture ran into serious financial trouble; at one stage Sir Bernard Lovell was even threatened with prosecution. During one major crisis he felt that the only thing to do was to drop everything for a few hours and play cricket — which he did. Speaking as a leg-break bowler myself, I fully understand!

When Sputnik 1 was launched, on 4 October 1957, the barely completed 'dish' at Jodrell Bank was the only radio telescope in the world outside Soviet Russia which was capable of tracking it. At once the Press, which had been antagonistic, changed its tune, and almost overnight Lovell was transformed from an eccentric spendthrift into a national hero. Finally, in 1960, Lord Nuffield came to the rescue and paid the outstanding debt of £50,000, solving all the main problems 'at a stroke'.

ions, and both have been objects of attention from radio astronomers.

Radio astronomy began in 1931, in a most unexpected manner. Karl Jansky, an American radio engineer of Czech descent, was instructed to study 'static' on behalf of his employers, the Bell Telephone Company. For this he built an improvised aerial, part of which was made from pieces of an old Ford car. Before long he realised that he was picking up signals which he could not identify, and after some time he was able to show that the 'radio noise' came from the Milky Way. We now know that the source was the star-cloud which lies in the direction of the centre of the Galaxy.

One would have expected Jansky to follow up his discovery, which was completely new and was obviously important. Yet he was strangely lethargic; he wrote a few papers in technical journals, which caused little interest, and then he turned his attention to other matters. Before World War II, the world's only intentional radio astronomer was Grote Reber, who constructed a dish-type aerial and managed to detect radio noise from the Sun.

It was only after 1945 that radio astronomy emerged as a true branch of science. Then, in the 1950s, came the 250ft dish at Jodrell Bank, the brainchild of Sir Bernard Lovell. It is fair to say that the Lovell Telescope, as it is now known, opened up a whole new avenue of research.

There are still people who fondly imagine (a) that a radio telescope picks up actual noise from the Sun and stars, (b) that you can look through it, and (c) that before long radio telescopes will make old-fashioned optical telescopes obsolete. All these ideas are wrong. Sound waves cannot travel in a vacuum, and there is practically no air above a height of a few miles above ground level; the 'noise' so often heard in broadcasts is produced entirely in the telescope's receiver, and is only one way of recording the incoming signals. Secondly, no radio telescope can produce a visible picture of the target object, and in general the end result is a trace on a graph. The cosmic radio maps which look so attractive are really in the nature of contour charts. And thirdly, radio astronomy and optical astronomy are complementary rather than being rivals.

The name radio telescope is misleading, because the instrument is more in the nature of a large aerial. With a metal dish, such as that at Jodrell Bank, the signals from space are collected and focused in the same way that an ordinary telescope collects and focuses light-waves, but instead of being sent into an eyepiece the signals go to a receiver which measures their intensities. Neither are all radio telescopes shaped like dishes. Many of them are quite different, and some have been compared with collections of barber's poles, but all have the same basic aim.

Early results were puzzling. The Sun turned out to be a radio source, as was only to be expected, and in 1955 long-wavelength emissions were also detected from Jupiter, but bright stars such as Sirius remained obstinately 'radio quiet'. Instead, apparently blank regions of the sky seemed to be powerful radio sources. It took years to find out what was actually happening. Radio emitters are of various kinds; in our Galaxy we have, for instance, the remnants of supernovæ such as the Crab Nebula, plus other decidedly exotic objects, while further away in the universe we have 'radio galaxies' and also the strange, super-luminous quasars, now generally believed to be the cores of very active systems.

In 1967 radio astronomers at Cambridge, headed by Sir Martin Ryle, were carrying out a special kind of investigation with a 'barber's pole' type of telescope when one member, Jocelyn Bell (now Dr Jocelyn Bell-Burnell) made an amazing discovery. She found a hitherto unknown radio source which seemed to be 'ticking' so quickly and so regularly that for a brief period it was genuinely believed that we might be picking up a signal sent out by some intelligent being far across the Galaxy. The attractive LGM or Little Green Men theory was soon disproved (though until it had been definitely ruled out, no statement was made to the Press — a wise precaution!) and we now know that what Jocelyn Bell had found was the first pulsar, a tiny, super-dense neutron star spinning rapidly round and sending out the curiously regular signals.

But would it be possible to detect intelligible signals from across space? If we are going to contact other civilisations, this seems to be the only possible way. Radio waves, remember, move at the same speed as light (186,000 miles per second), and nothing else can match that.

Preliminary attempts were made as long ago as 1961, with Project Ozma (readers of Baum will see here a connection with the famous

Odds against

There have been many forecasts about alien landings on Earth, but is it really likely that we will be visited in the foreseeable future? Not according to the bookmakers at Ladbroke's; in 1990 they offered odds of 100 to 1 against a visitation within the next ten years. (The same odds were offered against our sending a manned expedition to Mars during the next decade.) Significantly, both these eventualities were regarded as more probable than that of a British victor in the Men's Singles at Wimbledon, which was quoted at 200 to 1 against.

The 'Pulsar' radio telescope. Not all radio telescopes are dishes; this one, at Cambridge, was used to discover pulsars, and is a collection of poles. With me is Dr Antony Hewish (June 1968)

Eggen's star

Apart from the sun there are only eleven stars within ten light-years of us, and of these only two are visible with the naked eye: Alpha Centauri (4.3 light-years) and Sirius (8.7 light-years). All the others are dim red dwarfs apart from the Companion of Sirius, which is a white dwarf. Within twelve light-years we find five more naked-eye stars: Epsilon Eridani, 61 Cygni, Epsilon Indi, Procyon and Tau Ceti.

In 1976 O.J.Eggen reported the discovery of a red star in the constellation of the Sculptor which seemed to be very close. The announcement caused a great deal of interest, but it was then found that there had been a mistake; Eggen's Star is not particularly near.

Wizard of Oz!). Using the large radio telescope at Green Bank in West Virginia, later to collapse so suddenly and so dramatically, Dr Frank Drake and his colleagues listened out at a wavelength of 21.1cm, directing their attention to the two nearest solar-type stars, Tau Ceti and Epsilon Eridani. This wavelength was chosen because it is that of the signals sent out naturally by the clouds of cold hydrogen spread throughout the Galaxy, so that radio astronomers, wherever they might be, would be particularly interested in it.

No positive results were obtained, which was hardly a surprise. Similar though more elaborate experiments have been carried out since, and although the chances of success are low they are not nil. We have also sent out signals on our own account, using a mathematical code; after all, we did not invent mathematics, merely discovered it.

Of course, any radio contact with an alien civilisation would be somewhat leisurely. If we assume that Tau Ceti and Epsilon Eridani are the nearest stars likely to be attended by inhabited planets, it follows that a signal from Earth will take eleven years to arrive. If some obliging Cetian or Eridanian radio astronomer picks it up, and replies at once, his answer will take a further eleven years to reach us. Send out a message in 1990, and you may hope for a reply in 2012,

THE CRAB NEBULA

A comment from a famous astronomer, repeated so often that it has become almost hackneyed, is that 'there are two kinds of astronomy — the astronomy of the Crab Nebula, and the astronomy of everything else'. This may be overstating the case, but there is no doubt that the Crab Nebula in Taurus is one of the most interesting and informative objects in the sky.

Visually it looks like an undistinguished blurred patch; I can just see it in powerful binoculars, close to the third-magnitude star Alheka or Zeta Tauri not far from Orion. Telescopes show it easily, and with photographs it is seen to be a very complex structure, with entwined filaments. The nickname is due to the third Earl of Rosse, who drew it in the 1840s with the aid of his powerful telescope and said that it looked somewhat crablike.

We know what it is: the remnant of the supernova which was seen in 1054 by Chinese star-gazers. It became at least as bright as Venus, and was visible for many months before it faded from view. In 1731 a dim patch was seen by the British amateur John Bevis in just the position of the Chinese 'guest star'; it was rediscovered in 1758 by Charles Messier, who later compiled a famous catalogue of clusters and nebulæ and made it his No 1, while John Herschel said that it looked like 'a heap of sand'. Spectroscopes showed it to be gaseous, but it was not until our own century that astronomers were sure that it was truly the remnant of the 1054 supernova.

The Crab was one of the first radio sources to be identified with an optical object. Then, in 1969, astronomers at the Kitt Peak Observatory in Arizona detected a very faint, flashing object that we know to be a pulsar or neutron star; it spins round thirty times every second, and represents the actual core of the destroyed star.

When a Type II supernova explodes, material is ejected at a speed of at least 6,000 miles per second. This material rams into the interstellar gas and dust spread thinly through space, and this material is heated to several million degrees C, which is enough to generate X-rays. Eventually the ejected material slows down and the remnant ceases to glow, but with the Crab there has not been enough time for this to happen yet; its distance is 6,000 light-years, so that the outburst dates back for only 7,000 years or so from the present time. We can measure the expansion of the gas-cloud, and we also know that the pulsar's rate of spin is slowing down, though not by much — in fact, by approximately three thousand-millionths of a second per day.

Theoretical astronomers have reason to be grateful to the Crab Nebula. Perhaps our only regret is that the flare-up was seen in 1054 and not in 1990!

which, as some people have unkindly commented, is almost as slow as the Post Office.

Needless to say, we have absolutely no proof that there are radio astronomers in the systems of Tau Ceti, Epsilon Eridani or anywhere else. The only life which we definitely know to exist is that on our own Earth. Just how it arose is still very much of a mystery; most authorities conclude that it began in our warm oceans between 3,800 and 4,000 million years ago, though there is a minority view, championed by Sir Fred Hoyle, that it was brought to our world by way of a comet, and there has even been a suggestion by one of the world's greatest scientists, Francis Crick, that it was deliberately 'planted' here by a faraway alien race. But it is best to admit that we simply do not know.

Statistically, it seems reasonable to assume that life in the universe is likely to be widespread. Every time we have tried to set ourselves up on a pedestal we have been rudely humbled, and there is no reason to believe that we are of any special importance in the cosmos as a whole. Yet it may well be that there are no other intelligent races within many light-years of us. Our technological civilisation is young; if a visitor from Outer Space had arrived here a million years ago, which is not long on the astronomical time-scale, there would have been no scientists waiting to greet him or her (or it). Picture two lamps in a darkened room, each of which can be switched on at random for one second per week: what are the chances that they will shine at the same moment? Obviously not very great; and if a civilisation does not last for more than a limited period, there may be no others within our range at the present moment.

There is, too, a suggestion that Man may after all be unique, so that we represent the only advanced life-form in the entire universe. it is quite true that the emergence of life depends upon a number of special events, each of which is inherently improbable, but the fact remains that it has happened here, and I for one am reluctant to believe that it has not also happened elsewhere. Saying this is one thing, but proving it is quite another. If we had found any traces of life on Mars, the least unfriendly of our planetary neighbours, we would have proved the point — but as we did not, we must go on speculating.

We have already sent our first direct messengers into deep space. Pioneers 10 and 11, and Voyagers 1 and 2, will never come back to the Sun; as we have seen, they will simply go on and on until they are either destroyed by collision with some wandering body or are found and collected by another civilisation. Because of this possibility, there have been suggestions that we have been unwise, and that we are foolish to run the risk of drawing attention to ourselves. It is not a point of view that I support, but certainly there is a case to answer.

We have to admit, unfortunately, that the history of *Homo sapiens* is not a pleasant one. From the very beginning we have fought each other, tortured each other, and done our best to wipe ourselves out. The fact that we have not succeeded is only that we have lacked the necessary ability — up to now. Rome destroyed Carthage and drove a plough over its site; if Carthage had destroyed Rome instead, the

Eyesight test

If aliens approached the Earth in a space-craft, what evidence of man-made activity would they see with the naked eye by the time they reached the Moon, assuming that their eyes were about the same strength as ours? It has often been claimed that the only artificial structure which could be seen with the naked eye from that distance is the Great Wall of China, but this has been fully investigated by an expert, H.J.P. Arnold, who writes: 'The reality is that apart from a small renovated section near Pekin, the Wall is a frequently broken-down edifice. Far from being seen from space, it can often scarcely be seen on the ground. And although the human eye is extremely good at distinguishing lines from great distances, seeing the Wall from space is a physical impossibility.' And Neil Armstrong, the first man on the Moon, has said that there was absolutely no chance of seeing the Wall. Another legend bites the dust!

Returning echoes

A space-probe from another world was described in 1974 in a book by a Scottish writer, Duncan Lunan. He based this remarkable conclusion by analysing the results of radio signals sent out from Earth in 1927–8; the returning echoes were delayed in a way which showed that they were being diverted by a satellite which had been sent out by the inhabitants of a planet in the system of the star Izar (Epsilon Boötis), which is 200 times as luminous as the Sun and is 150 light-years away. Apparently the Izarians were calling for help.

It seems rather a cumbersome method of sending an alarm signal, and since no further evidence has come to hand we must assume either that the space-probe has gone away, or that it was never there!

history of the world would have been different, but I would still be sitting at my desk writing this book. The change came in 1945, with the fall of the first atom bomb on Japan. Since then, 'improved' nuclear weapons have been not only planned, but actually produced, so that if we blundered into another major conflict we could reduce the entire Earth to a sterile, radioactive waste. Can we be confident that if we had indeed found a civilisation on Mars, our contacts with it would have been peaceful? Similarly, would any alien race reaching our world come in a spirit of friendship?

We are certainly safe from attack within the Solar System, because there is nobody else here, but we cannot definitely say the same about space travellers from other systems. Once they know about us, they may decide to take us over — so the pessimists say; and this is why they have been anxious to destroy our Pioneers and Voyagers before they have had the chance to betray our presence. This seems absurd, if only because none of our probes can come within range of another star for thousands of years, but in any case we have already given ourselves away simply because of our radio and TV broadcasts.

We cannot travel in time; only characters such as Flash Gordon, Dr Who and Lord Darth Vader can do that. But we can do the next best thing; we can look back in time, and, as I have pointed out, our view of the universe beyond the Solar System is bound to be very out of date. To give just a few examples: Alpha Centauri is just over 4 light-years away, Sirius 8½, Altair 16½, Vega 26, Capella 42, Castor 46, Regulus 85, and so on. If you look at Castor now in 1991, you are actually seeing it as it used to be in 1945. Now stretch the imagination beyond reasonable limits, and visualise an astronomer in the Castor system who is equipped with a telescope powerful enough to see the Earth in detail. When he looks at us, what does he see? A world at war, with Britain and Germany at each other's throats. An equally well-equipped astronomer in the system of Polaris, 680 light-years away, would have a grandstand view of the Crusades.

Our own broadcasts began around 1925, so that by now they have travelled some sixty-five light-years into space. From Sirius, Altair and Castor we are 'radio noisy', but from Regulus we are still 'radio quiet', because not until the year 2010 will our first broadcasts penetrate out as far as that. When they do, a Regulan radio astronomer might realise that signals have suddenly started to come from the neighbourhood of an insignificant yellow star far across the Galaxy, and he would become aware of the existence of a technological civilisation there. Therefore, say the critics, the damage has already been done. Any aliens within about sixty light-years of us, equipped with sufficiently sensitive equipment, will know just where we are.

This does not dismay me in the least, for one excellent reason. As yet we have no idea of how to achieve interstellar travel, because rockets are much too slow, and devices such as space-arks are likely to remain in the realm of science fiction. If we are to reach the stars, we must await some fundamental breakthrough. There may well be races which have achieved it, and know how to cross great distances, but in this case they will be much too enlightened to feel like coming to attack us. They would more probably be able to teach us a great deal.

So when I look up at Sirius, Vega, Altair, Castor, Regulus and the rest I am not at all alarmed. Also, let me give one suggestion which is purely my own but which is not, I feel, outrageous. If a space-craft from another world did land here, its occupants would know how to talk to us, because they would have taken the trouble to listen to our broadcasts, and learn our language, before arriving.

I hope it will happen one day. Time will tell.

The Goldstone Radio telescope, one of the main receivers for the signals from deep-space probes

Island Universes

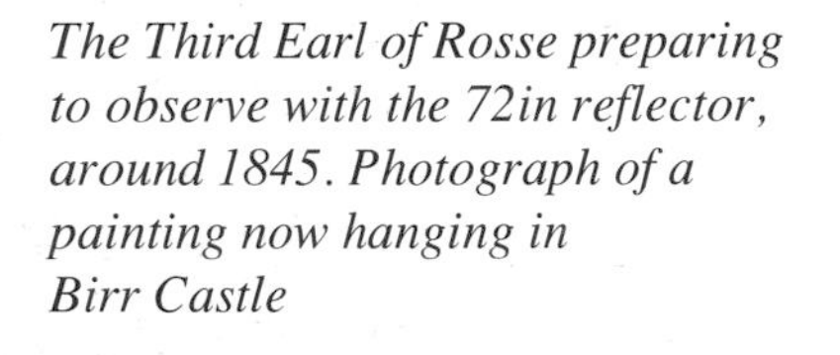

The Third Earl of Rosse preparing to observe with the 72in reflector, around 1845. Photograph of a painting now hanging in Birr Castle

Can you name the strangest telescope ever built? There are no doubt several claimants to the title, but a strong contender would be the 72in reflector built in 1845 by the third Earl of Rosse, and set up at his country seat in Birr Castle. This is almost exactly in the middle of Ireland (the nearest large town, Athlone, is some miles away). The weather there is not very good, to put it mildly; Lord Rosse had no skilled helpers; he had to make his own mirror, not of glass but of metal, and there were no engineering works upon which he could call. Everything had to be done single-handed, even to the casting of the mirror, which involved making a full-scale forge. The

Above:
The Horse's Head Nebula in Orion. The dark nebula bears a distinct resemblance to the shape of a knight in chess (Mount Wilson)

Right:
The globular cluster in Hercules (M13), the brightest globular cluster visible from Britain. Taken with a 10in reflector (Bernard Abrams)

Seyfert Galaxy NGC 1068. Photographed by E. Giraud with the New Technology Telescope at La Silla, 1989 (European Southern Observatory)

Railway nuisance

The Rev Dr Romney Robinson was the third Director of the Armagh Observatory. His régime lasted from 1823 to 1882, and he was mainly responsible for putting the Observatory on a sound footing.

He was always ready to assert himself, as was shown by his battles with the authorities over the Newry-Enniskillen Railway. He complained that passing trains shook the sensitive telescopes and other instruments, and was determined to prevent new lines from coming any closer; so far as he was concerned, the Newry and Armagh Railway Corporation was an unmitigated nuisance, though it is probably fair to add that the railmen must have felt the same about Dr Robinson. In 1856 a petition was presented to Parliament, and resulted in an Act prohibiting the trains to go closer than 700 yards of the Observatory without prior permission from the Governors, which, of course, would never be given. Robinson made a sinister entry in his notebook about the later branch line: 'It will not come within the limit, I believe, but I shall watch them carefully.'

scene when the mirror was cast was graphically described by an onlooker, the Rev Romney Robinson, Director of the Armagh Observatory in what is now Northern Ireland:

> The sublime beauty can never be forgotten by those who were so fortunate as to be present. Above, the sky, crowded with stars and illuminated by a most brilliant moon, seemed to look down auspiciously on their work. Below, the furnaces poured out huge columns of nearly monochromatic yellow flame, and the ignited crucibles during their passage through the air were fountains of red light, producing on the towers of the castle and the foliage of the trees, such accidents of colour and shade as might almost transport fancy to the planets of a contrasted double star. Nor was the perfect order and arrangement of everything less striking; each possible contingency had been foreseen, each detail carefully rehearsed; and the workmen executed their orders with a silent and unerring obedience worthy of the calm and provident self-possession in which they were given.

When completed, the telescope was mounted between two massive stone walls. It could swing only to a limited extent to either side of the north-south line, which meant that Lord Rosse had to rely upon the Earth's rotation to bring target objects into view. There was no finder, and the telescope was simply 'aimed' by eye. Moreover, there was no mechanical drive. The observer had to perch atop the telescope in what might seem to be a precarious position, and it gave the impression of being decidedly unsafe. Not that Lord Rosse minded; he was also meticulous about his observing dress, and when going to the telescope he never failed to put on his top hat.

It might be thought that this telescope, often nicknamed 'the Leviathan', would be nothing more than a curiosity. On the contrary, it proved to be one of the most significant instruments in the whole history of astronomy, and in its day it was just as much of a leader as the Mount Wilson 100in reflector and the Palomar 200in were in later times. Lord Rosse was a genius in his own way; alone and unaided, he built what was then much the most powerful telescope in the world, and used it to make fundamental discoveries. Had he been a lesser man, he would have tried to make his telescope more convenient and more elaborate — and he would have failed. It was by accepting the telescope's limitations that he managed to succeed.

He paid particular attention to the faint patches of light which we call nebulæ. Many were known, and probably the most famous list of them had been drawn up in 1781 by the French astronomer Charles Messier, who was totally uninterested in them but who was persistently misled by them during his searches for new comets. Altogether, Messier's catalogue included over a hundred objects, including most of the famous clusters and nebulæ; thus the Crab Nebula is M1, the Orion Nebula M42, the Pleiades M45, the Andromeda Spiral M31, and so on. Sir William Herschel had added thousands of new nebular objects, and other astronomers had joined in, so that by the time Lord Rosse completed his 'Leviathan' he had many targets to examine. On the other hand, nobody knew just what they were.

In some cases, of course, there could be no doubt. The Pleiades group is an obvious star-cluster; so too is Præsepe in Cancer (Messier's No 44) and also the Hyades round Aldebaran, which have no Messier number, presumably because they could not possibly be confused with a comet. Then there were objects such as M13 in Hercules, which were regular in form and are today known as globular clusters. A relatively small telescope will show that globular clusters are resolvable into stars, at least in their outer portions.

M42, in Orion's Sword, was different. Stars could be seen in it — or in front of it? — but the overall impression was that of shining gas. There were also dark patches, which seemed to be material which was not being lit up by any embedded star. The comparison between 'gaseous nebulæ' of this sort and 'starry nebulæ' such as M31 in Andromeda was striking.

Lord Rosse began to survey the nebulæ, and almost at once he made a discovery which proved to be more important than he realised at the time. Many of the starry nebulæ were spiral in form, like Catherine wheels. In some cases, as with M51 in the constellation of the Hunting Dogs, the spirals were face-on; with other objects they lay at a narrow angle, as with the brightest of all the starry nebulæ, M31 in Andromeda, which is almost edgewise-on to us, so that the full beauty of the spiral is lost.

In the mid-nineteenth century there was no telescope in the world, other than the 'Leviathan', which was powerful enough to show the spirals, and of course photography was not advanced enough to record them, but Lord Rosse had the spirit of a true scientist; anything he discovered was made available to all, and astronomers from all

One galaxy only?

In the Bakerian Lecture, delivered in 1888, Sir Norman Lockyer — founder of the periodical *Nature*, and also of the observatory at Sidmouth in Devon which bears his name — claimed that all self-luminous bodies in space were made up of swarms of meteorites, either densely or loosely packed, and radiating because of the frictional heat produced when the meteorites collided with each other. This idea never became popular, but the famous scientific historian Agnes Clerke wrote in her book *The System of the Stars* that 'there is but one island universe — that within whose boundaries our temporal lot is cast, and from whose shores we gaze wistfully into infinitude. Dismissing, then, the grandiose but misleading notion that nebulæ are systems of equal rank with our Galaxy, we may turn our attention to the problems presented by their situation within it.'

Nicknames

Some galaxies have nicknames: the Sombrero Hat, the Whirlpool, the Pinwheel and so on. One nearby dwarf system, very small indeed by the standards of galaxies, has been given the derisive nickname of Snickers.

over the world were free to come and make use of the great telescope. By the time the 72in had come to the end of the main part of its brilliant career and had been superseded by more modern-type instruments (which did not really happen until the 1880s), the existence of many spirals had been confirmed.

Not all the starry nebulæ were spirals. Some were elliptical, some spherical, some irregular. But could there be any real distinction between the various types of nebulæ — that is to say, were they all members of our Galaxy, or were some of them independent systems in their own right? That was what nobody could decide.

Measuring the distance of a star is difficult enough, but measuring

INSIDE THE HUNTER'S SWORD

The most famous of all the gaseous nebulæ is in the Sword of Orion. It is easily visible with the naked eye as a hazy patch, and binoculars show it very well. Yet strangely there is no mention of it, as a nebula, before it was described in 1610 by an otherwise obscure astronomer named Nicholas Peiresc. The ancients said nothing about it; in 1603 Johann Bayer merely recorded a star in its position, and even Tycho Brahe and Galileo missed it. We can hardly suppose that it chose to brighten just at the exact moment when telescopes were invented, but it is certainly a minor mystery. In Messier's catalogue it is No 42.

The Orion Nebula is 1,500 light-years away. It is made up of very thin gas (mainly hydrogen, with a considerable amount of helium) with 'dust'. David Allen has pointed out that if you could take a core sample right through it, with the core being 1in across, you would not collect enough material to balance the weight of a pound coin — and yet the diameter of the Nebula is 30 light-years, so that if it were centred on our Sun we would find that bright stars such as Sirius and Altair lay inside it.

Nebulæ are illuminated by stars lying in or very near them. In the case of M42 there are the four components of the multiple Theta Orionis, nicknamed the 'Trapezium' because of their arrangement. They lie on the near side of the cloud as seen from Earth, and are probably no more than 100,000 years old; they are so hot that they make the nebular material 'fluoresce' and send out a certain amount of light on its own account. Nebulæ of this sort are termed emission nebulæ, as against those which shine purely by reflection.

Obviously we can see only the near edge of the Nebula, but we have nevertheless been able to probe inside it. Ordinary light cannot pass through the nebular material, but infra-red can do so, and powerful infra-red sources deep inside M42 have been located. One of these was found by the American astronomers Eric Becklin and Gerry Neugebauer, and is therefore known generally as BN. It was once believed to be a very young star which was still condensing and was not yet hot enough to radiate in visible light, but we now know it to be a fully-fledged giant, only a few tens of thousands of years old, and with a surface temperature of 20,000°C. It is running through its evolution much more quickly than the Sun, and given enough time it would 'bore a hole' through the surrounding nebulosity and make itself visible, but it will never have the chance, so that so far as we are concerned it must remain hidden until it dies. It is only one of several strong infra-red sources inside the Nebula. Moreover, we know that we are seeing a stellar nursery in which fresh stars are being born; there are numbers of variable stars which have not lived for long enough to join the Main Sequence and are still 'flickering' irregularly.

Dark patches can also be seen in M42, and some way away from the main glow, near Alnitak (the southernmost member of the Hunter's Belt) there is a starless patch which has been nicknamed the Horse's Head because of its resemblance to the head of a knight in chess. There is no difference between a dark nebula and a bright one apart from the lack of a suitable illuminating star. For all we know, there may be no such star on the far side of M42, so that if we were looking at it from a different vantage point it might appear as inky as the Coal Sack in the Southern Cross.

Altogether the Orion Nebula is a fascinating place. We know, too, that it is only the visible part of a huge 'molecular cloud' which covers most of the constellation of the Hunter.

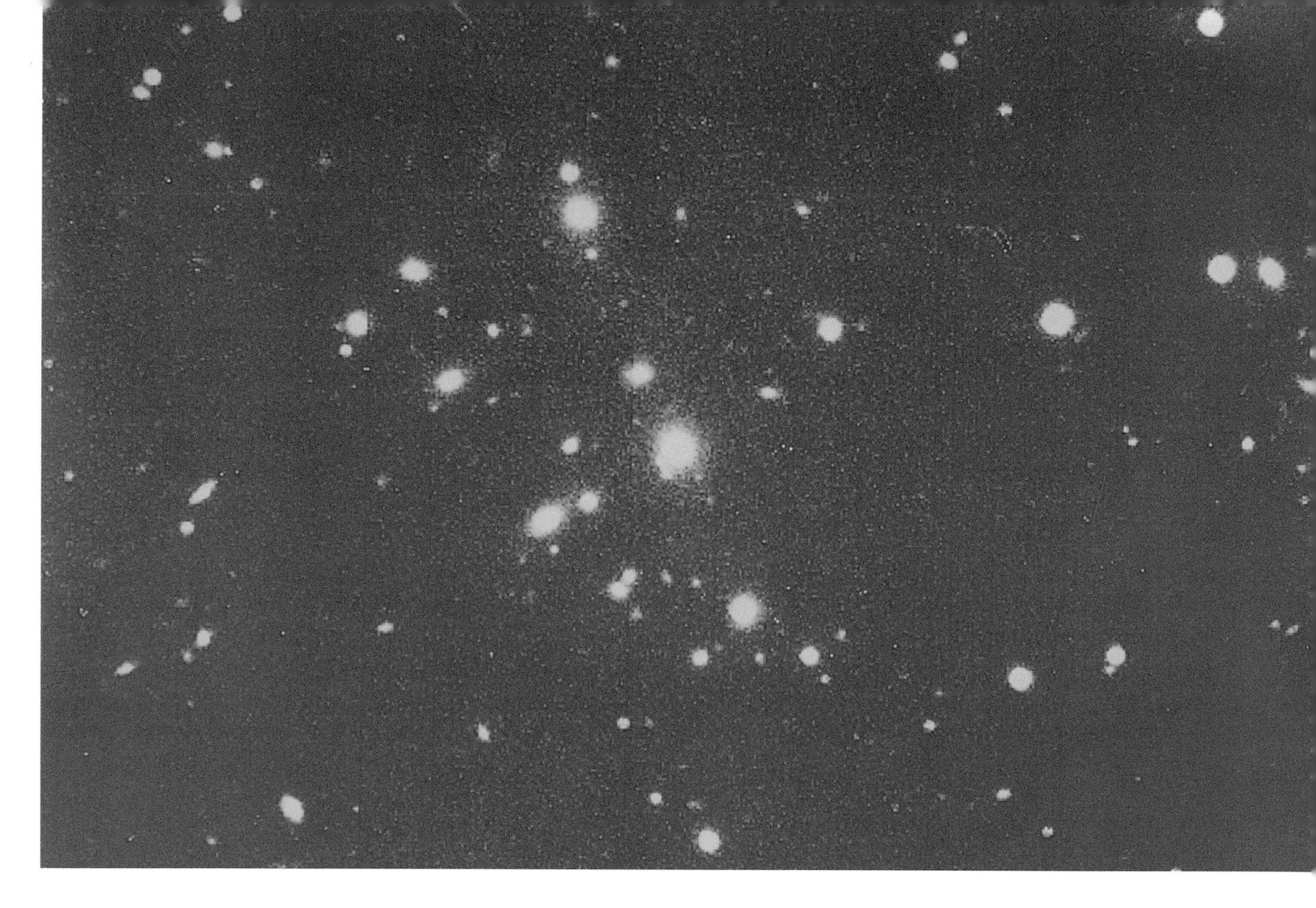

Remote cluster of galaxies, in Hydra. Photographed by E. Giraud, with the New Technology Telescope at La Silla, 1989 (European Southern Observatory)

the distance to a nebula is much more difficult still, because the annual parallax shift is too small to be detected. William Herschel had suggested that the starry nebulæ might be true galaxies or 'island universes', but he could give no proof, and neither could Lord Rosse. There had to be a fundamental breakthrough, and it finally came many years later in a most unexpected way.

In the far south of the sky are the two patches known as the Magellanic Clouds, because they were described by the Portuguese navigator Ferdinand Magellan during his voyage round the world in 1519. Magellan was certainly not the first to see them, because they are conspicuous naked-eye objects, but neither he nor anyone else at the time had the slightest idea of their nature.

In 1912 Miss Henrietta Leavitt, of Harvard Observatory, was studying photographs of the Small Cloud which had been taken from the Observatory's southern station in Peru. In the Cloud she found some stars of the type known as Cepheid variables, after the brightest member of the class (Delta Cephei, in the northern part of the sky). Most stars shine steadily over long periods, but the variable stars do not; they brighten and fade relatively quickly, and with Cepheids the periods of variation — that is to say, the intervals between successive maxima — range from a few days to several weeks. It was also known that the Cepheids, unlike other variable stars, are absolutely regular in their behaviour, so that they repeat the same cycle time after time.

Miss Leavitt found that in the Small Cloud of Magellan, the brighter Cepheids were always those with the longer periods of

Unusual astronomer

John Goodricke was one of the most unusual astronomers who has ever lived. He was born in 1764, and spent most of his short life in York; he was deaf and dumb, but there was nothing the matter with either his eyesight or his brain. He discovered the fluctuations of two very important variable stars, Delta Cephei and Beta Lyrae, and correctly explained the fadings of Algol as being due to an eclipsing companion. He died in York at the early age of twenty-one.

variation; there was a definite relationship. For example, a Cepheid with a period of eleven days would be brighter than a Cepheid with a period of nine days. But it was clear that the Cloud is so far away that to all intents and purposes the stars in it can be assumed to be the same distance from us, just as it is normally good enough to say that Victoria Station and Piccadilly Circus are the same distance from Australia. It followed that the Cepheids in the Cloud which looked the

Changing horizons

The Parkes 210ft dish radio telescope, completed in 1963, is still one of the largest and best in the world; it lies in New South Wales, not far from the optical observatory at Siding Spring. When it was found that the 1963 occultation of the radio source 3C–273 by the Moon would not be accessible from England, and so could not be studied from Jodrell Bank, the Parkes team decided to go ahead. To their consternation, they found that 3C–273 would be so low over the horizon that the dish could not be tipped down far enough to reach it. Drastic action was called for. Local trees were hacked down, and it was also necessary to dismantle part of the telescope's gearing mechanism, with teeth being filed off. Even this was barely sufficient, and the astronomers could do no more than observe the very end of the occultation. Luckily this was enough; they obtained an accurate position for the radio source — and the first of all quasars had been tracked down.

brightest really were the most luminous. Working along these lines, Miss Leavitt was able to formulate a rule according to which a Cepheid would 'give away' its luminosity simply by its behaviour.

If this were true of the Cepheids in the Small Cloud, then there could be little doubt that it was also true of Cepheids elsewhere, and at once astronomers were provided with an invaluable set of what could be called standard candles. Moreover, Cepheids are very

The 210ft radio telescope at Parkes, New South Wales

The great debate

In 1920 came the famous Great Debate in America between Heber D. Curtis (who believed the spirals to be true galaxies rather than parts of our Milky Way) and Harlow Shapley (who did not). Curtis proved to be right. On the other hand, Shapley's estimate of the size of our own Galaxy was much better than Curtis', so that we may agree that the result of the Great Debate was an honourable draw.

powerful stars, often thousands of times more luminous than the Sun, so that they can be seen over immense distances.

Next in the story came Harlow Shapley, who set out to follow a career in journalism and became an astronomer instead. In 1917 he was working at the Mount Wilson Observatory in California, and since he was particularly interested in the size of the Galaxy he realised that Miss Leavitt's discovery would be of tremendous help. He looked carefully at the globular clusters, and realised that they are not spread evenly all over the sky. This could only be because we were having a lop-sided view. If the globular clusters lay around the edges of the main Galaxy, it followed that we were ourselves a long way from the centre of the system.

Shapley took photographs of the globular clusters, and in them he found Cepheid variables. He could find out the distances of the Cepheids merely by observing them, and hence he could also find out the distances of the clusters in which they lay. Before long he was able to give a remarkably good estimate of the size of the Galaxy. We now know that it has an overall diameter of around 100,000 light-years, and that the Sun, together with the Earth and the other planets, lies slightly less than 30,000 light-years from the centre of galactic nucleus, which we can never actually see — there is too much 'dust' in the way — but which we know to lie beyond the star-clouds of Sagittarius.

THE DOPPLER EFFECT

Most people have noticed the Doppler effect at one time or another, even though they may not have known what it means. The name comes from the Austrian physicist Christian Doppler, who first drew attention to it as long ago as 1842. It was afterwards extended by the French scientist Hippolyte Fizeau, and the French still refer to it as the Doppler-Fizeau effect.

Listen to the sound of an ambulance or police car which is coming toward you, with its siren blaring. As the vehicle approaches, the note of the siren will be high-pitched, because more sound-waves per second reach your ear than would be the case if the vehicle were standing still, and the wavelength is effectively shortened. As soon as the vehicle has passed by and has started to recede, fewer sound-waves per second enter your ear, so that the wavelength is apparently lengthened and the note of the siren drops. This is the original Doppler effect.

The same sort of thing happens with light (as Fizeau pointed out). With an approaching light-source the wavelength is shortened, and the object appears 'too blue', while with a receding source the wavelength is lengthened, making the object look 'too red'. Generally the actual colour-change is much too slight to be noticed (do not expect a yellow traffic-light to turn blue if you race toward it!) but it shows up when spectroscopes are used.

Remember that the spectrum of an ordinary star consists of a rainbow or continuous background crossed by dark lines. Under laboratory conditions these lines have fixed positions, but with an approaching object they are shifted over to the short-wave or blue end of the rainbow, while with a receding object the shift is to the red. The amount of the shift is a key to the velocity of approach or recession — the greater the shift, the greater the speed. Consider, for instance, Rigel in Orion and Vega in Lyra. In the spectrum of Rigel there is a slight red shift, from which we can tell that the star is moving away from us at a rate of 13 miles per second; Vega has a slight blue shift, showing that it is approaching at 8¾ miles per second.

The fact that all the galaxies beyond our own Local Group show red shifts in their spectra indicates that the entire universe is expanding — unless, of course, the shifts are not pure Doppler effects, as some eminent astronomers believe.

Yet Shapley was wrong in one important respect, because he believed that the spirals and other starry nebulæ were minor features of our own Galaxy. It was only in 1923 that the next great breakthrough came. It was again made at the Mount Wilson Observatory, but this time with the great 100in reflector, which was incomparably the most powerful in the world and which had been brought into use a few years earlier. Using it, Edwin Hubble looked at the spirals, and found that they too contained Cepheid variables.

One system which came under his scrutiny was M31 in Andromeda, which is the only spiral clearly visible with the naked eye from northern latitudes. Hubble's Cepheids indicated that it must be approximately 900,000 light-years away, in which case it had to be well beyond the boundary of our Milky Way, and would rank as a separate galaxy or 'island universe'. Subsequently, Hubble reduced his estimate of the distance of M31 to 750,000 light-years. Both these early values were much too low, but at least it was clear that our Galaxy was only one of many; Cepheids were soon found in other systems also, and the old term of 'spiral nebulæ' became more or less obsolete, though it is still used occasionally.

Meanwhile, another interesting fact was becoming evident. At about the same time that Miss Leavitt was studying her pictures of the Small Cloud, Vesto Melvin Slipher, at the Lowell Observatory in Arizona, had been photographing the spectra of some of the nebulous objects. (Though Percival Lowell had founded the observatory mainly to study Mars, a great deal of other work was also carried on there; Slipher became Director when Lowell died in 1916.) It was found that apart from a few cases, the so-called starry nebulæ showed red shifts. If these were due to Doppler effects, it followed that the objects were moving away from us.

Hubble followed this up as soon as he had established the true nature of the galaxies, and again he found that there was a definite relationship; the further away a galaxy was, the faster it was moving away. In other words the entire universe seemed to be expanding, with every galaxy receding from every other galaxy.

There were a few exceptions, notably the Andromeda Spiral, but all these exceptions were nearby by cosmical standards, and gradually it emerged that galaxies cluster in groups. Our Local Group consists of a few large systems, including M31, plus the medium-sized Clouds of Magellan and more than two dozen much smaller members. Other clusters of galaxies are much more populous, and the Local Group is by no means exceptional. Neither are we in any privileged position. The best analogy is that of a balloon upon which inkspots have been marked; blow up the balloon, and each spot will recede from each other spot. Of course, we can fix the centre of the balloon, and we cannot fix the centre of the universe, but the analogy is still a useful one.

Hubble and his colleague, Milton Humason, found that the speeds of recession were staggeringly high. Then, in 1952, came another revelation, due to the German-born, American-by-adoption astronomer Walter Baade. By then the 200in Hale reflector at

Unwelcome resident

The telescope used to take the photographs of the Small Magellanic Cloud which were used by Miss Leavitt in her studies of short-period variables was a fine 13in refractor, with an object-glass made by Alvan Clark, America's leading telescope expert. It was then at Arequipa in Peru, the outstation of Harvard Observatory. When Arequipa was closed down, the telescope was taken to the Boyden Observatory at Bloemfontein, in South Africa, where it is still in regular use.

A cobra has made his home under the floorboards of the Observatory. The Director, Dr Alan Jarrett, told me casually that 'He lives there, but we don't bother about him; he leaves us alone, and we leave him alone. No problem — live and let live.' Being a natural coward I was not entirely reassured, particularly on being told that a visiting astronomer, entering the dome ahead of his escort, had been greeted with a rearing 10ft snake, but I am delighted to say that I have not yet personally made the cobra's acquaintance. It is something which I am fully prepared to forego.

Peculiar Galaxy ISSO 060-1G 26. Photographed by N. Taranghi with the New Technology Telescope at La Silla (European Southern Observatory)

Palomar had been completed, and was just as great an improvement on the Mount Wilson telescope as the 100in had been over Rosse's Leviathan. Using it, Baade was able to show that there had been a major error in the Cepheid scale. What Hubble had not known, and could not have been expected to know, was that there are two brands of Cepheids, Type I and Type II. Those of Type I are twice as luminous as those of Type II, and so far as the galaxies were concerned astronomers had chosen wrongly. The Cepheids which had been used for estimating the distances of the galaxies were twice as powerful, and hence twice as remote, as had been supposed. In one short paper delivered to the Royal Astronomical Society in London, Baade calmly doubled the size of the universe. (I well remember the stunned reaction of the audience.)

The identification of quasars, in 1963, made the situation even more difficult to appreciate. How could one believe in a relatively small object, no larger than the Solar System, shining with a power equal to over a hundred whole galaxies — bearing in mind that a typical galaxy may contain a hundred thousand million stars? Moreover, the red shifts in the spectra of quasars indicated enormous speeds of recession, amounting in some cases to over 90 per cent of the velocity of light.

These results appeared to impose a limitation upon the distance to which we could see, no matter how powerful our telescopes. It was said that if the rule of 'the further, the faster' holds good, then we must eventually come to a region in which a galaxy or a quasar will be receding at the full speed of light. We will naturally be unable to see it at all, and we will have come to the boundary of the observable universe. This is not the same as saying that we have actually probed to the limit of the universe itself, but only that we have penetrated as far as is theoretically possible from our own particular vantage point. We may be getting somewhere near it. Hubble's relationship gives the critical distance at somewhere between 15,000 million and 20,000 million light-years, and a quasar discovered in 1989 has a red shift which places it at least 14,000 million light-years away.

The weakness in the whole argument is that we depend entirely upon the assumption that the red shifts in the spectra of galaxies (and quasars) are pure Doppler effects. If so, then our distance measurements are unquestionably of the right order. But just suppose that we are not dealing with pure Doppler effects, and that the red shifts are due, in part at least, to a different cause? This would cause a dramatic upheaval in all our theories, and it is certainly the view of some astronomers, one of whom is Sir Fred Hoyle, who has made fundamental contributions to the study of stellar evolution. Another is Dr Halton C. Arp, formerly of the Mount Wilson Observatory and now working at the Max Planck Institute in Germany, and it is Arp who has provided the first real evidence.

What he has done is to check the relationships between quasars and galaxies which lie side by side in the sky, forming groups, lines or clusters. There are often visible 'bridges' of material between them, in which case there are presumably real physical connections. Yet Arp has found that there are many cases of objects which are quite obviously associated, and have completely different red shifts. Either the alignments are due to sheer chance, or else the red shifts are not pure Doppler effects, making all our distance measurements unreliable.

Traditional astronomers have viewed Arp's results with utter consternation. This is understandable enough; if he is right, then most

Honest mistakes

The Dutch astronomer Adriaan van Maanen emigrated to the United States after graduating, and in 1912 joined the staff of the Mount Wilson Observatory. He carried out some excellent work, but he also claimed that his photographs showed distinct movements of the stars in the outer parts of the Andromeda Spiral, in which case the Spiral could not possibly be as far away as had been indicated by Edwin Hubble's studies of the short-period variables. Hubble challenged van Maanen's results, and it was eventually found that there had been errors in the measurements. The mistakes were perfectly honest, but the situation was not improved by the fact that Hubble and van Maanen disliked each other intensely.

Quasar OQ-471. The quasar (arrowed) is one of the most remote objects known to astronomers (Palomar Observatory)

other people are wrong. Instead of being remote and super-luminous, quasars could be comparatively close, perhaps even members of the Local Group. Arp himself believes that they are minor features ejected from young galaxies.

In any case, there can be little doubt about the significance of Arp's results, and we must remember that as well as being one of the most experienced observers of the present day he has also had the advantage of using some of the world's best telescopes. I do not pretend to be a cosmologist, but it seems fairly evident that something is badly wrong. We could well be on the threshold of a revolution which will take us back possibly not to Square One, but certainly to Square Two.

This may not be a bad thing. Arp has been strongly criticised for casting doubt upon the conventional picture of the universe, but this is the fate of all pioneers; less than a century ago, the idea of spiral systems as external galaxies had been more or less relegated to the limbo of forgotten things. And it is no longer heretical to suggest that the idea of super-powerful, super-remote quasars may be equally obsolete in a few years' time.

Astronomical wager

I am not usually a betting man, but in 1990 I agreed a wager with an old friend who is one of Britain's leading astronomers. I claim that by the end of the decade it will have been shown that Halton Arp's ideas are right, and that we will have to revise all our ideas about the distances of quasars. The debt is due to be paid on 1 January 2001, the first day of the new century. The loser will pay the winner one bottle of whisky — 1990 vintage!

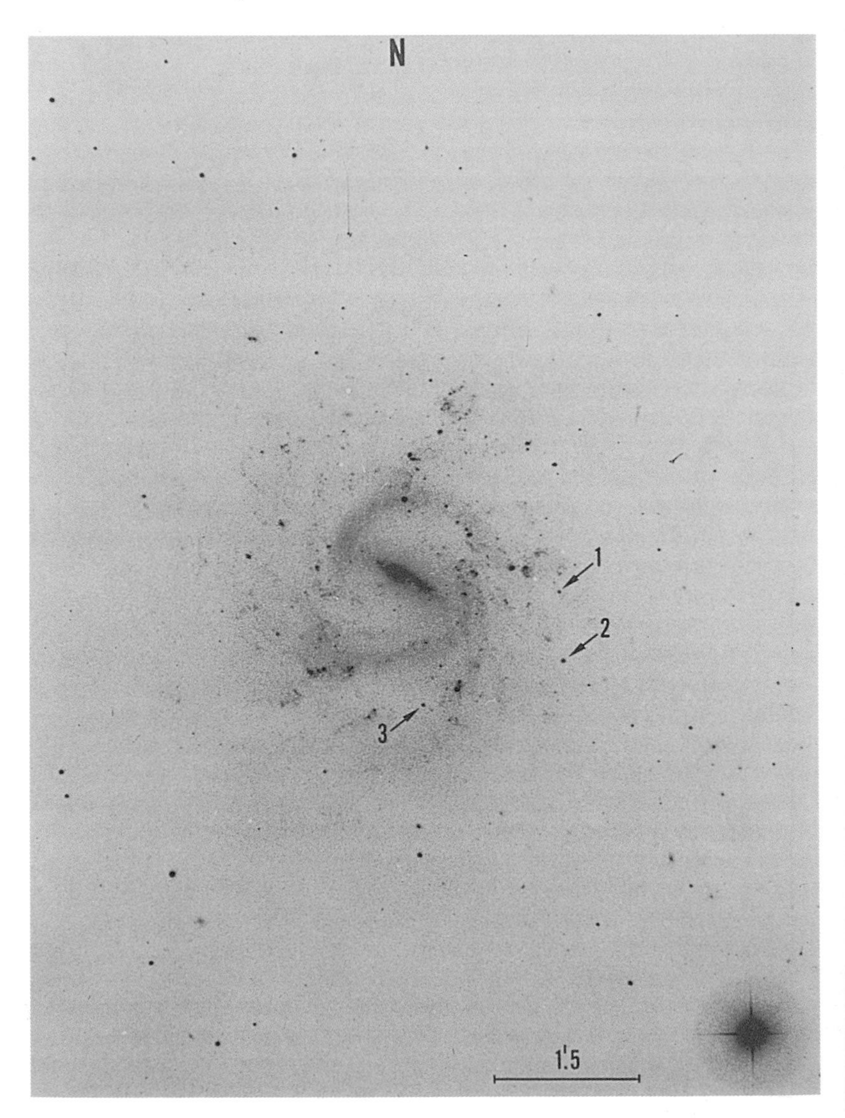

Quasars and galaxies , apparently associated but with different red shifts (H. C. Arp)

The End of Time

Many years ago, when I was a teenager, I had the pleasure of meeting the great novelist H. G. Wells. He impressed me enormously, and I then made it my business to read all those of his books which I had not previously found. I devoured them all, but, predictably, there were three which I found particularly compelling: *The Time Machine, The War of the Worlds,* and *The First Men in the Moon.*

The Time Machine was Wells' first novel, and arguably his best (though everyone will have personal views about this). As you may remember, the hero of the story builds a machine which can send him either into the past or into the future. He explores a world in which Man has become decadent, with the graceful but ineffective Eloi battling against the dark, evil Morlocks; he penetrates much further into the future, when Man has vanished altogether from the Earth and nothing is left but a desolate landscape illuminated by a dying Sun. It is all quite fascinating, and the story-telling is superb, but has it any scientific basis?

Wells, unlike his great predecessor Jules Verne, was a qualified scientist — but also unlike Verne, he made no attempt to keep to proper science; for example, he knew quite well that the lunar world which he described in *The First Men in the Moon* is pure fantasy. Time travel, of course, has always been a favourite science-fiction theme, and has been exploited by many writers of lesser calibre than Wells; and in a way, travel into the future is possible, though not by any Wellsian mechanical device.

It all comes back, as usual, to Einstein's theory of relativity. Go on a long journey, moving at near the speed of light, and your time-scale will slow down in relation to that of a companion who stays at home. This gives us the famous Twin Paradox. One twin sets off on a trip to a distant star, moving at, say, 99 per cent of the velocity of light. If the journey there and back takes him something of the order of twenty years, he will return home to find that a far longer period has elapsed on Earth, so that his twin is either old or dead. If he stays away for long enough, travelling fast enough, he could even come home to shake hands with his great-great-grandchildren. This sounds absurd,

Time-travel

My favourite time-travel story was written by, of all people, Stephen Leacock. In his *Nonsense Novels* we find 'The Man in Asbestos', in which the narrator sends himself to sleep for many hundreds of years simply by forcing himself to read the leading column of the New York *Times* through from beginning to end. He wakes up to find a world where everything has been standardised; people are dressed in uniform asbestos suits, education is performed by means of surgical operation, and international disputes are settled by means of a slot-machine (which, come to think of it, isn't such a bad idea after all).

but there are experimental proofs that what we call the 'time-dilation effect' really does happen.

Cosmic rays provided the first clue. These are not genuine rays at all, but high-speed particles bombarding us all the time from outer space, coming in from all directions. When they crash into the upper air, they break both themselves and the air particles; new particles called mu-mesons are created. These mu-mesons decay so quickly that they ought not to last long enough to reach the Earth's surface. Yet they do — because they are travelling so fast that their time-scale, relative to ours, is slowed down. Modern experiments have shown the same effect. A very accurate watch flown round the world at high speed will show a slight but measurable difference from a watch which remains in the laboratory.

At speeds which astronauts can manage at the moment, the time-dilation effect is absolutely negligible; before it becomes important, you have to work up to something like 90 per cent of the velocity of light, which is hopelessly beyond us. As yet we cannot achieve even 1/10,000 of the velocity of light. Obviously we cannot predict what we will be able to do eventually, and it would certainly be intriguing to set off upon an interstellar journey with the knowledge that we would return to the world centuries or even thousands of years hence, but there is one vital point to bear in mind; it would be a one-way trip only. Having returned to the world of AD10,000 we could never go back to AD1990.

Orgone energy

Wilhelm Reich, who lived from 1897 to 1957, is regarded as one of the founders of modern psychoanalysis. He studied under Sigmund Freud, and held important posts in Viennese psychoanalytical institutions. He believed in what he called orgone energy, which permeates all Nature; it is blue, and is responsible for lightning, the colour of the sea, and the luminosity of the tails of glow-worms, as well as human sexual activity. According to Reich, galaxies are formed when two streams of orgone energy rush together, producing matter which eventually condenses into stars, planets, and human life. Even by the standards of Reich's profession, this seems rather extreme!

Even if we admit the reality of the time-dilation effect, as we are bound to do, there is a much greater question-mark against travel into the past. Like the White Queen in *Through the Looking-Glass,* I am quite ready to believe in several impossible things before breakfast each day; but in my view, at least, travelling backward in time is one of the few things which must be regarded as genuinely impossible. There can be no conceivable way of getting into Wells' time machine, or Dr Who's Tardis, and shooting oneself back to see Julius Cæsar landing on the coast of Britain, King Canute addressing the waves, or, for that matter, W.G. hitting a boundary. Quite apart from other problems, a traveller into the past could presumably influence the future, and the implications of this are obvious!

Time may be regarded as a dimension. For example, a book lies on a chair, and its position can be defined in terms of length, breadth and height, but we must also include time, because the book may not have been there a few minutes ago. Extra dimensions come into the realm of abstruse mathematics, but time as a dimension is at least something which can be described in everyday language.

It is also possible to give some idea of the time-scale of the universe. If current theories are right, then the universe in its present form began at some period between 15,000 million and 20,000 million years ago. The age of the Earth is known, fairly reliably, to be between 4½ and 5 thousand million years. Life has existed here for a large fraction of that period, but intelligent life is very new.

If we represent the total age of the Earth by scaling it down to one year, we can work out a time-scale of evolution. For the first part of our 'year', the world was lifeless. The first primitive organisms appeared in the warm seas during early May. They were single-celled, and rather resembled our modern algæ, but they were also our direct ancestors.

Little happened throughout the summer of our time-scale, but by late October there were more elaborate forms of marine life instead of the original single-celled organisms. It was not until 20 November that the first real fishes appeared in the seas, and not until 30 November that there were the first forays on to dry land. At that period the atmosphere was not the same as it is now, because it contained much more carbon dioxide and much less free oxygen. If a modern man were transported there, he would promptly choke.

Launch of Jules Verne's Columbiad. The space-gun principle was followed by Jules Verne in his classic novel From the Earth to the Moon. *This is a woodcut from the original edition of the book*

Reptiles came next. By the end of the first week in December they were dominant, and we had the fearsome dinosaurs as well as other dinosaurs which were harmless and vegetarian. By 15 December the earliest mammals had made their entry, though they could not gain a real foothold until the dinosaurs disappeared, which may have been suddenly, around 65,000,000 years ago (as the cosmic collision supporters believe) or more gradually, over a period of up to a million years or so. Only at 5pm on 31 December do we come to the first 'hominids', evidenced by footprints left on a fall of volcanic ash on Laetoli Plain in Kenya, and the arrival of *Homo sapiens* was delayed until only an hour before midnight on 31 December. This means that on our cosmic scale, the Battle of Hastings happened less than a second ago.

All this we can accept, because there is firm scientific evidence in

WHAT KILLED THE DINOSAURS?

Dinosaurs have always caught the popular imagination. Certainly these remarkable reptiles ruled the world for a very long time — far longer than Man has done. Yet around 65,000,000 years ago they vanished, and today we know them only as fossils.

What was the cause? The 'death of the dinosaurs' has been the subject of endless discussion, and even now we do not pretend to know the answer. One thing we can rule out at once is competition from other living creatures. Though some of the dinosaurs were small and harmless, and many were actually vegetarian, there were others which were the most fearsome creatures ever seen on Earth, and no enemies could have stood up against them (apart, of course, from other dinosaurs), so that we must look elsewhere for a reason. Note, by the way, that it was not only the dinosaurs which made their exit at this time — the end of the geological period known as the Cretaceous — because many other species of animals and plants vanished as well; moreover, there have been earlier extinctions on a similarly grand scale, one of which occurred at the end of the Permian Period, around 230,000,000 years ago. For the moment, however, let us confine ourselves to looking at the various theories about the dinosaurs.

1 Purely natural 'exhaustion'. This assumes that the race was simply worn out, so that reproduction ceased without any external interference. In this case the process would have taken many thousands of years.

2 A sudden change in the output of the Sun, altering the Earth's climate so markedly that the dinosaurs, which were not noted for their intelligence, were unable to cope with the new conditions.

3 The passage of the Solar System through a 'cosmic cloud' which temporarily dimmed the sunlight, plunging the Earth into a long period of cold and gloom which again would have been fatal to the dinosaurs.

4 A sudden increase in harmful radiation from space, which might have been due to any one of a number of causes. This would at least explain why so many other species died out at the same time as the dinosaurs.

5 A tremendous impact by a meteorite, an asteroid or a comet, which would have caused a dramatic change in climate. Support for the meteorite idea has been forthcoming from investigations of rocks laid down at around the same period, and which contain a surprising amount of the rare substance iridium — which could well be meteoritic. Alternatively, it has been proposed that there is a massive planet moving round the Sun in a very eccentric orbit, and that at intervals it disrupts the Oort cloud of comets and sends missiles hurtling into the inner part of the Solar System, in which case the Earth would have been subjected to a violent bombardment. The hypothetical planet has even been given a name — Nemesis — and supporters of the theory claim that it comes back every 26,000,000 years or so, in which case it is overdue! However, I feel that we should not become alarmed, and the whole scheme is so speculative that one can only be sceptical.

Everything really hinges on whether the extinction of the dinosaurs was sudden or gradual. Again we cannot be sure, because a period of a million years or so is very 'long' by our everyday time-scale but very 'short' to a geologist. However, we may be sure of one thing: the dinosaurs were gone long before Man appeared on the face of the Earth. We can forget the idea of Mr and Mrs BC locking up the cave and putting the iguanodon out for the night.

support of it, but we cannot really appreciate it, because it is too vast for our limited brains. Moreover, there is another brick wall ahead. We say glibly that the universe, in its present form, began at least 15,000 million years ago; what happened before that?

Assuming that everything — space, time, matter, the whole universe — began with a 'Big Bang', we find that theory can take us back almost to the beginning. We can trace events back to 10^{-43} seconds after the Big Bang (10^{-43} is a convenient mathematical shorthand, meaning 10 divided by 1 followed by 43 zeros). Before this, theory can help us no more, because all presently acknowledged scientific laws break down, and we are reduced to nothing more than speculation. But can we be sure that there really was a Big Bang?

Not all astronomers believe so. In the late 1940s a group at Cambridge University, headed by Herman Bondi and Thomas Gold (later joined by Sir Fred Hoyle) put forward an alternative picture, according to which the universe has always existed and will exist for ever; there was no Big Bang and no moment of creation. They admitted that the universe is expanding, so that galaxies will pass over the edge of the observable region and will be cut off from all contact with us, but they assumed that new galaxies are being formed spontaneously out of nothingness in the form of hydrogen atoms. The rate of creation would be too slow to be detectable, but it would be enough. The 'steady state' theorists did not attempt to explain how new matter was spontaneously born, but, after all, to suppose that hydrogen atoms appear like rabbits out of a hat is no more fantastic than supposing that the entire universe appeared suddenly in the same way.

Unlike most speculative theories, this one could be tested. It would mean that the universe would always look much the same as it does now. If we could come back in, say, a million million years, we would see the same numbers of galaxies as we do today, and they would be spread around in the same manner, even though they would not be the same galaxies. It followed that the distant regions of the universe, available for our inspection, must be similar to the regions nearer at hand, since if, for example, we look at a galaxy 10,000 million light-years away, we are seeing it as it used to be 10,000 million years ago. Also at Cambridge, Sir Martin Ryle and his team of radio astronomers carried out a systematic survey, and proved that the distribution of remote galaxies is not the same as with galaxies closer to us. Therefore the universe is not in a steady state, and the whole theory is wrong. The coup de grâce was given by the discovery of background radiation, coming from all directions, which we can assume to be the last remnants of the Big Bang.

We must next decide whether the Big Bang was an unique event. At the moment it is generally agreed that the groups of galaxies are racing away from each other, and that the further apart they are, the faster they are going. (This follows the conventional view, not Halton Arp's.) What we do not know is whether this expansion will continue indefinitely, so that at last all the galaxies will lose touch with each other, or whether it is merely the present state of affairs. Everything

Fossil hoax

Professor Johann Beringer, of Würzburg, was keenly interested in fossils, and collected them enthusiastically. Unfortunately he made the mistake of offering rewards to his students, paying a a considerable sum for each fossil brought in, with the inevitable result that the students started manufacturing their own fossils. Before long the luckless professor had accumulated a vast collection, but the crunch came when he discovered that one of the fossils had his own name carved on it. By that time he had written a large book, *Lithographic Würceburgensis,* giving the full results of his research. When he realised that he had been the victim of a hoax, he spent the rest of his career in trying to buy up and destroy all the published copies of his book — with scant success.

End of the Greenwich Time Signal. Professor Alec Boksenberg, Director of the Royal Greenwich Observatory, officially hands over the 'six pips' time signal at a ceremony in the Time Room at Herstmonceux Castle. The signal is now transmitted from the BBC. This ceremony, in 1990, really marked the end of Herstmonceux as an astronomical centre

depends upon the average density of matter in the universe. If there is enough of it, then the galaxies will eventually stop, turn, and begin to come together again, so that after an immense period of around 80,000 million years there will be a new Big Bang; the universe will be re-born, and the cycle will start all over again. On this 'cyclic' theory we can trace the story back as far as we like, with periodical Big Bangs, so that again we can dispense with the need for a single act of creation.

This theory too can be tested, but as yet we do not have enough evidence in our hands, because we still do not know how much matter the universe contains. If it exceeds the critical value (about 3 atoms per cubic metre) then the galaxies cannot escape. If the density is below this critical value, then the galaxies can never be drawn back, and the expansion will never stop. On the cyclic theory, the universe resembles a clock which is being regularly re-wound; otherwise, it may be likened to a clock which has been wound only once, and will finally run down.

Most people will instinctively opt for the first theory, because the alternative is depressing — a universe which is terminally ill, and which will end with isolated corpses of galaxies from which all energy has departed and all signs of life have become extinct. Yet at the moment we have to concede that the amount of matter we can actually see is much too little to pull the galaxies back. It is inadequate by a large margin, even if we take into account all the observable stars, galaxies and interstellar material. On the other hand there is also very strong evidence that there is a tremendous amount of 'missing mass' which we cannot account for. It may be locked up in black holes; it may be explained by myriads of low-mass stars which we cannot see because they are too feeble; there may be forms of matter which are quite undetectable by our present techniques — we simply do not know. But we are sure that it exists, and the 'missing mass' problem is perhaps the most important outstanding problem of modern astronomy.

All in all, science can take us a long way, but not as far as the basic fundamentals. As has been stressed many times, we have never

Leap seconds

The year 1990 began one second late. Just before midnight on 31 December 1989 all clocks were put forward by one second, so that the last broadcast time signal of the year consisted of seven pips instead of six. Though the Earth's rotation is being slowed by tidal friction, and on average each day is 0.00000002 second longer than its predecessor, there are also slight random variations due to factors such as movements in the Earth's core, changing ocean currents, and tidal patterns; this necessitates occasional 'leap seconds', because our modern atomic clocks are much better timekeepers than the Earth itself. In the Devonian geological period, 350,000,000 years ago, the day was only twenty-two hours long.

discussed the 'origin' of the universe from a purely scientific point of view. We have to start with *something,* and from this we can trace the story through to the present day. What defeats us is how the 'something' came into existence in the first place. And here we wander away from scientific fact into different fields which I admit that I am not qualified to discuss, though I have the consolation that nobody else is much better off.

The only people who claim to be able to give full answers are the Biblical Fundamentalists, who believe that the account given in Genesis is literally true. God produced the world in seven days, and there is no need to go into further details of exactly how this was achieved.

Of course, this is only one of the many accounts of the Creation; turn to any of the other major religions of the world and you will find a different story, but the underlying principle is exactly the same. A great deal depends upon dating, and James Ussher, Archbishop of Armagh in Northern Ireland, was in absolutely no doubt about the matter. Between 1650 and 1654 he published an exact chronology, in which he stated that the Earth was created at nine o'clock in the morning of 21 October, 4004BC. He derived this date by the simple method of adding up the ages of the patriarchs and making various other calculations of the same kind which were, no doubt, extremely scholarly, but which, like W. S. Gilbert's flowers that bloom in the spring, have nothing to do with the case. The Ussher date was widely supported by Church officials until well into the second half of the nineteenth century!

Satellite laser ranger at Herstmonceux Castle. This sends out a laser beam which enables the distance of a satellite to be determined with remarkable accuracy. When the Royal Greenwich Observatory left Herstmonceux, in 1990, the laser ranger remained in place there

Of course, a recent origin for the Earth can be disproved simply by examining the record of fossils, which take us back several hundreds of millions of years at least. So far as fossils were concerned, the Biblical Fundamentalists either ignored them or else dismissed them as frauds — though one leading Victorian theologian, Edmund Gosse, sidestepped the problem very neatly by claiming that although the world had indeed been created in 4004BC, fossils had been produced by God at the same time for the set purpose of deceiving future palæontologists.

The evolution of life was another controversial topic. Darwin's

ECHOES OF THE BIG BANG

If the universe came into existence with a Big Bang, at least 15 thousand million years ago, it must initially have been very hot. During the first tenth of a thousandth of a second following the moment of creation the temperature was probably of the order of a million million degrees C. At once the infant universe began to spread out in all directions, cooling as it did so, while the wavelength of the radiations increased. After only three minutes the temperature had fallen to a few thousand million degrees C, and the nuclei of hydrogen and helium atoms began to form. These, like all atomic nuclei, had positive electrical charges. At this stage there was still so much energy around that the positive nuclei were unable to join up with the negatively charged electrons, and the universe was opaque to radiation, simply because no radiation could travel very far before being absorbed by interactions either with matter or with other radiations.

This state of affairs lasted for about half a million years, but then there was a change. Matter and radiation became 'de-coupled'; that is to say, the nuclei and the electrons were at last able to link up, so that the electrons were no longer able to interact with the radiation, and so far as radiations were concerned the universe became transparent.

As the universe continued to expand, the background radiation went on increasing in wavelength until it had reached the microwave part of the electromagnetic spectrum. (Remember that microwaves extend between about 1mm and 1m in wavelength.) There was also the reduction in temperature, which by now has dropped to a mere 3°C above absolute zero.

Scientifically, 'temperature' is not the same as what we always call 'heat'; it depends upon the speed at which the various atoms and electrons move around — the faster the speeds, the higher the temperatures. Consider, for instance, the solar corona, which is the Sun's outer atmosphere and is made up of extremely rarefied gas. The atoms are moving out so quickly that the 'temperature' is over a million degrees C, but there is very little 'heat', because there are too few atoms. The best everyday analogy is to compare a firework sparkler with a glowing poker. Every spark is 'white-hot', but contains so little mass that there is no danger in holding the firework in one's hand — while I for one would be very reluctant to pick up the end of a poker which has been left in the fire.

The slower the velocities, the lower the temperature, and there must be a stage when the movements have stopped, so that the temperature can fall no further; we have come to absolute zero, which is equivalent to -273°C. This is also taken as the starting-point for the Kelvin scale, named after the great physicist Lord Kelvin, so that 0°C is the same as 273 Kelvins, usually written 273K. Therefore, the temperature of the present background radiation is written as 3K.

The existence of a 3K microwave background had been predicted in the early 1940s, but nobody had taken much notice, and nothing more was done until 1964, when a most curious episode took place. At the Bell Telephone Laboratories in New Jersey two experimenters, Penzias and Wilson, were testing a microwave receiver used for communications with artificial satellites when they found that they were picking up signals in the microwave region which they could not explain. Puzzled, they consulted the team at Princeton University led by Robert Dicke. Quite unknown to them, Dicke had been carrying out calculations which led him back to the idea of a 3K background — and this was precisely what Penzias and Wilson had found!

It has been said, with good reason, that the discovery of the microwave background was the most important development in astronomy for the past fifty years. There seems little doubt that it represents the last traces of the Big Bang which happened so long ago.

theory, published in 1859, was hotly attacked by the Church, though it is worth stressing that Darwin never suggested that Man is descended from monkeys, as his opponents usually claim; all Darwin pointed out was that men and monkeys have a common ancestry, which is quite true. (It is amazing that in recent times 'Creationism' has had something of a revival, particularly in America, where in some States it is still regarded as heretical to teach 'Darwinism'.)

Though it is easy to dispose of theories such as Ussher's, it is perfectly true that when it comes to explaining just how the universe was born the scientists are no better off than the Fundamentalists. Whichever way we turn, we are up against a blank wall. There are two clear alternatives:

1 The universe was created at a set moment in time.
2 The universe had no beginning; it has always existed.

In case (1), we have to ask 'What happened before the creation?' The only answer is to say that nothing happened, because 'time' itself had not started; but this is no real help.

With (2), we have to visualise a period of time which had no beginning. This is something which we are incapable of doing, because the concept of infinity is too much for us.

Therefore it is best to say frankly that we are beaten. Whether any other race, in some other part of the Galaxy, has been able to solve the problem is something else which we do not know, but looking up into the sky on a starlit night must give us food for thought. It is surely inconceivable that in all this host of worlds there is not another civilisation which is older and wiser than ours, so that it might be able to guide us if there were any means of contact.

This also brings home the fact that there is order and method in the universe, so that presumably there is an underlying power controlling it. I am not referring to the dogma of any particular religion; and bearing in mind that most of the wars on Earth — both past and present — have been triggered off by religious agitation, surely none of our present faiths can be said to be beneficial to humanity as a whole (to find something better, we have to go back to Akhenaten). If life began and ended here, so that we die when our bodies do, the entire process would be pointless — and whatever else the universe may be, it is not pointless. But this is leading into metaphysics, so let us return to the main theme, that of the evolution of the universe itself.

Whether or not there were a single moment of creation, the one inescapable fact is that we exist. We have a future, but on the physical plane it is a limited one. The Earth itself cannot hope to survive the fierce blast of radiation from the Sun during its red giant stage, so that if humanity still survives when the first danger signals become apparent there will have to be a concentrated effort to see what can be done. Moving the Earth into a different orbit is totally beyond our powers as yet, but our technological civilisation is very new, and the Sun will not change much for at least two thousand million years, so that we have plenty of time to spare. On the other hand, it may be

Pigeon trouble

During their experiments on behalf of the Bell Telephone Corporation, Penzias and Wilson were baffled by the unidentifiable radiation which they were picking up. The antenna which they were using was shaped like a horn, and the local pigeons used to make use of it, leaving their mark in no uncertain manner. Could pigeon droppings be sending out the mysterious signals? It seemed possible, and so Penzias and Wilson painstakingly cleaned off all the droppings and did their best to keep the pigeons away. The signals were still there. As they came from distant parts of the universe, this was hardly surprising.

Cosmic explorer

It is generally thought that immediately after the Big Bang there was a period of rapid expansion – 'the inflationary period'. If the universe had then been completely 'smooth', it is not easy to see how our present very far from smooth universe could have developed. The background radiation discovered by Penzias and Wilson seemed to be absolutely uniform in all directions, and this presented theorists with a grave problem. Then, to their great relief, space research came to their rescue. In 1992 a satellite, COBE – the Cosmic Background Explorer – found that the 3K radiation was not absolutely smooth after all; there were tiny irregularities. All was well once more, and the theorists breathed sighs of relief.

The Kuiper Airborne Observatory. This is merely an aircraft specially equipped to carry a telescope to great altitudes

essential to evacuate the Earth altogether, if only because the Sun will change from a luminous red giant into a very small, feeble white dwarf. At least there is a strong chance that we will have mastered the art of interstellar travel long before the crisis is upon us.

This is fantasy, just as the idea of reaching the Moon would have been regarded as fantasy in the time of the Crusades, but even if we could save ourselves by moving or abandoning the Earth there are even more pressing problems ahead. The Galaxy cannot last for ever, and neither can the other galaxies we can see. Eventually, all of them must die. On the cyclic theory there will be another Big Bang, though long before that happens the closing-in of the galaxies, many of them dead, will have snuffed out all life anywhere in the universe. If the cycle then begins anew, then there may well be the same sequence of events, with new worlds and new civilisations which will also end with yet another Big Bang — and so on *ad infinitum*. Otherwise the universe will come to an end in its present form, and we will be left with nothing but a cold, dead void.

In this case, will 'time' also come to an end? We are in just as much difficulty as we were in considering the creation, and like everyone else, I am incapable of suggesting an answer.

Although we do not know how the universe was born, or how (or if) it will die, we have at last started to understand many facets of it. We have found, too, that the Earth is peculiarly suited to us inasmuch as it has the right kind of temperature, the right sort of air, and a plentiful supply of water. It could be our home for thousands of millions of years to come, but whether or not it remains friendly and fertile is for us to decide. We could so easily ruin it, either by intent — that is to say, by a final destructive war which would turn the whole globe into a radioactive waste — or by negligence, pollution, depletion of our reserves, or disease following upon over-population. Very probably the next century or two will determine which road we will follow.

Meanwhile, we have our opportunity. Go outdoors on the next dark, clear night, and look up at the stars. Then, like me, you will realise that we live in a wonderful universe, and you will, without doubt, share my passion for astronomy.

Gamow's insight

George Gamow, who first predicted the existence of the microwave background, was an extraordinary character. He was Russian, but spent much of his life in the United States. On one occasion he attended an important conference dealing with cosmology, and slept through most of it, snoring loudly. Subsequently he sent a letter to the Chairman in which he gave a masterly solution of the main problem under discussion, leading one participant to comment wryly that Gamow, asleep, could think more clearly than other astronomers who were wide awake!

Glossary

Absolute magnitude The apparent magnitude that a star would have if it could be observed from a standard distance of 10 parsecs, or 32.6 light-years.
Albedo The reflecting power of a planet or other non-luminous body. A perfect reflector would have an albedo of 100 per cent. (The average albedo of the Moon is only about 7 per cent!)
Altazimuth mount A type of telescope mount upon which the instrument can be swung freely in any direction. This means that to guide it involves two movements, altitude (up or down) and azimuth (east or west). Thanks to modern computers, most new telescopes are now mounted in this way.
Ångström unit One hundred-millionth part of a centimetre. The wavelength of visible light ranges between about 3900 Å (violet) to 7500 Å (red). The unit was named after the nineteenth-century Swedish physicist Anders Ångström.
Aphelion The position in the orbit of a planet or other body when it is furthest from the Sun.
Apogee The position of a body in motion round the Earth when it is at its greatest distance from the Earth.
Apollo asteroids Asteroids whose orbits cross that of the Earth.
Ashen Light The faint luminosity of the 'night' side of Venus. It is generally attributed to electrical phenomena in the planet's upper atmosphere.
Asteroids (minor planets) Small planetary bodies, most of which move round the Sun in the region between the orbits of Mars and Jupiter, though there are some which depart from the main swarm. Only one (Ceres) is as much as 500 miles in diameter, and only one (Vesta) is ever visible with the naked eye.
Astrology The pseudo-science which attempts to link human characters and destinies with the positions of the Sun, Moon and planets in the sky. It has no scientific basis, and has long since been completely discredited.
Astronautics The science of space research.
Astronomical twilight The period between sunset and the time when the Sun has dropped 18° below the horizon.
Astronomical unit The distance between the Earth and the Sun: 92,957,209 miles (in round figures, 93,000,000 miles).
Astrophysics The branch of modern astronomy which deals with the physics and chemistry of the stars.
Aten asteroids Asteroids whose orbits lie mainly inside that of the Earth.
Atom The smallest unit of a chemical element which retains its own particular character. There are ninety-two known naturally occurring elements, and it is safe to say that none has been overlooked.
Aurora Auroræ are the lovely Northern Lights (Aurora Borealis) and Southern Lights (Aurora Australis). They occur in the Earth's atmosphere, and are due to electrified particles sent out by the Sun.
Azimuth The angular bearing of an object in the sky, measured from north (0°) through east (90°), south (180°) and west (270°) back to north (360° or 0°).
Baily's beads Brilliant points seen along the Moon's edge at a total solar eclipse, just before and just after totality. They are caused by the sunlight shining through valleys on the limb of the Moon, between mountainous regions.
Barycentre The centre of gravity of the Earth-Moon system. Because the Earth is eighty-one times as massive as the Moon, the barycentre lies deep inside the Earth's globe.
Big Bang theory The theory that the universe came into existence at one definite moment in time, between 15,000 million and 20,000 million years ago.
Binary star A star made up of two components, genuinely associated, and moving round their common centre of gravity.
Binoculars A pair of binoculars consists of two small refractors joined together. The magnification and the aperture in millimetres determine the type of binoculars: thus with 7x50 binoculars, the magnification is 7 and each object-glass is 50mm in diameter.
Black drop An appearance seen during a transit of Venus. As the planet moves on to the Sun's disk it seems to draw a strip of blackness after it; this persists until the

transit is well under way. It is due to the atmosphere of Venus.

Black dwarf A dead star, which has exhausted all its energy.

Black hole The region round an old, collapsed giant star where the pull of gravity is so strong that not even light can escape from it.

Bode's Law An interesting relationship linking the distances of the planets from the Sun. It is probably no more than coincidence.

Bok globule A small dark object, probably a proto-star, seen against a background of stars or a gaseous nebula. They are named in honour of Bart J. Bok, the Dutch astronomer who first drew attention to them.

Bolide A brilliant meteor, which may explode as it descends through the Earth's atmosphere.

Canals, Martian Straight, artificial-looking lines on Mars drawn by many astronomers from around 1877. They have no real existence.

Captured rotation (synchronous rotation) If the axial rotation period of a body is equal to its revolution period, the rotation is said to be captured. Tidal forces over the ages are responsible.

Cassegrain reflector A type of reflecting telescope in which the main mirror has a central hole; the light from the target object is reflected back through this hole by a smaller convex secondary mirror.

Cassini division The main division in Saturn's ring system.

Celestial sphere An imaginary sphere surrounding the Earth, whose centre is the same as that of the Earth. The celestial sphere is divided into two hemispheres by the celestial equator.

Cepheid A short-period variable star. The periods range from a few days to a few weeks, and are absolutely regular. There is a definite law linking the star's period with its real luminosity; the longer the period, the more luminous the star.

Ceres The largest and first-discovered of the asteroids. It was found by G. Piazzi in 1801, and has a diameter of 584 miles. It is never visible with the naked eye.

Charge Coupled Device (CCD) An electronic device, far more effective than a photographic plate. Most large telescopes are now used with CCDs.

Chromosphere That part of the Sun's atmosphere which lies immediately above the bright surface or photosphere.

Chronometer A very accurate form of timekeeper.

Circumpolar star A star which never sets, but merely circles the celestial pole, remaining above the horizon all the time.

Clusters, Stellar Genuine groups of stars. Open clusters have no particular form; globular clusters, as their name suggests, are spherical systems, containing in some cases over a million stars.

Coal sack The famous dark nebula in the Southern Cross.

Comet A member of the Solar System. A comet consists of a small icy nucleus, which when warmed by the Sun produces a head or coma by evaporation; sometimes there may be a tail or tails. Most comets move in very eccentric orbits.

Conjunction (a) The apparent close approach of two bodies in the sky, due to line of sight effects. (b) A planet is said to be at superior conjunction when it is on the far side of the Sun with respect to the Earth. Mercury and Venus can also come to inferior conjunction, when they lie almost between the Sun and the Earth, and are 'new'. (With exact alignment, a transit across the Sun's disk occurs.)

Constellation A pattern of stars. Since the stars are at very different distances from us, a constellation is nothing more than a line of sight effect, and the names are quite arbitrary.

Corona The outermost part of the Sun's atmosphere, made up of very tenuous gas.

Cosmic rays High-speed particles reaching the Earth from outer space.

Cosmic year The name often given to the Sun's period of revolution round the centre of the Galaxy: approximately 225,000,000 years.

Cosmology The study of the universe as a whole, its nature, and the relations between its various parts.

Culmination The maximum altitude of a celestial body above the horizon. The Sun, of course, culminates at noon.

Day The period taken for the Earth to spin once on its axis. A sidereal day (rotation with respect to the stars) has a length of 23hr 56min 4·091sec; a solar day (rotation with respect to the Sun) 24hr 3min 56·555sec — because the Sun has a daily eastward motion against the stars.

Degree of arc 1/360 of a full circle. Each degree is divided into 60 minutes of arc, and each minute into 60 seconds of arc.

Density The amount of matter in a unit volume of space. For most purposes the density of water is taken as 1.

Dichotomy The exact half-phase of Mercury, Venus or the Moon.

Direct motion Bodies which move in orbit, or rotate upon their axis, in the same sense as the Earth. Those which move or rotate in the opposite sense have **retrograde** motion.

Diurnal rotation The apparent daily rotation of the sky from east to west, due to the real rotation of the Earth from west to east.

Doppler effect The apparent change in wavelength of light (or sound) due to the motion of the source relative to the observer. With an approaching object, the wavelength is apparently shortened, giving a blue shift; with a receding object, the wavelength is apparently lengthened, giving a red shift.

Double star A star made up of two components. In some cases the appearance is due to chance lining-up, but in most cases the double stars are **binary** or physically associated pairs.

Earthshine The luminosity of the night hemisphere of the Moon, seen when the Moon is at the crescent stage. It is due to light reflected on to the Moon from the Earth.

Eclipse (a) Lunar: the passage of the Moon through the shadow cast by the Earth. (b) Solar: the temporary blotting-out of the Sun's bright disk by the interposition of the Moon.
Eclipsing binary A star made up of two components, one brighter than the other. When one component passes in front of the other, the light seems to fade as seen from the Earth. The best-known eclipsing binary is Algol in Perseus.
Ecliptic The apparent yearly path of the Sun against the stars, passing through the constellations of the Zodiac.
Electromagnetic spectrum The full range of wavelengths, from radio waves through microwaves, infra-red, visible light, ultra-violet, X-rays and gamma-rays.
Electron A fundamental particle carrying unit negative charge; it makes up part of an atom.
Element A substance which cannot be chemically split into simpler substances.
Elongation The apparent angular distance of a planet or comet from the Sun, or of a satellite from its primary planet.
Ephemeris A table showing the predicted positions of a celestial body, such as a planet or comet.
Equator, Celestial The projection of the Earth's equator on to the celestial sphere. It divides the sky into two hemispheres.
Equatorial mount A type of mounting in which the telescope is set upon an axis which is parallel to the axis of the Earth. When the telescope is moved to follow the movement of a celestial body, the change in altitude will therefore be automatic.
Equinox The places where the celestial equator cuts the ecliptic. The Sun crosses the equator twice, around 22 March (moving south to north) and 22 September (moving north to south). The spring equinox is often termed the First Point of Aries.
Escape velocity The minimum velocity which an object must have in order to escape from the surface of a planet, or other body, without being given any extra impetus.
Exosphere The outermost part of the Earth's atmosphere.
Eyepiece The lens, or combination of lenses, placed at the eye-end of a telescope. Its rôle is to magnify the image formed by the object-glass (of a refractor) or main mirror (of a reflector).
Finder A small telescope attached to a larger one.
Fireball A very brilliant meteor.
Flares, Solar Brilliant outbreaks in the Sun's atmosphere, usually associated with active sunspot groups. They send out electrified particles and short-wave radiations.
Focal length The distance between the centre of a mirror (or lens) and its focus.
Focus The point where rays of light meet after being converged by a lens or mirror.
Free fall The normal state of motion of an object in space under the gravitational influence of a central body. Thus the Earth is in free fall round the Sun.
Galaxies Independent star-systems, often made up of more than a hundred thousand million stars.
Galaxy, The The star-system of which our Sun is a member. It is often, though not accurately, called the Milky Way system.
Galilean satellites The four large satellites of Jupiter: Io, Europa, Ganymede and Callisto.
Gegenschein A very faint glow in the sky, exactly opposite to the Sun, and very difficult to see. Its English name is the **Counterglow.**
Gibbous phase The phase of the Moon when between half and full. Mercury and Venus also appear gibbous, and so, to a much lesser extent, can Mars.
Gravitation The force of attraction which exists between all particles of matter in the universe.
Great circle A circle on the surface of a sphere (such as the Earth) whose plane passes through the centre of the sphere. Thus a great circle will divide the sphere into two equal parts.
Greenwich mean time The local time at Greenwich. It is used as the standard throughout the world.
Greenwich Meridian The line of longitude which passes through a certain point in the Old Royal Observatory at Greenwhich.
Halley's Comet The only bright comet which returns regularly. It has a mean period of seventy-six years. It was last at perihelion in 1986, and will be back once more in 2061.
Halo The spherical-shaped star cloud round the main Galaxy.
Harvest Moon In the northern hemisphere, the full moon closest to the autumnal equinox (around 22 September).
Heliacal rising The rising of a celestial body at the same time as sunrise. However, it is commonly taken to mean the date when the body first becomes visible in the dawn sky.
Hertzsprung-Russell Diagram (often known as an HR Diagram). A diagram in which the stars are plotted according to their spectral types and their absolute magnitudes, (see page 123).
Horizon The great circle on the celestial sphere which is everywhere 90° from the observer's overhead point or zenith.
Hunter's Moon The first full moon following Harvest Moon.
Immersion The entry of a celestial object into occultation or eclipse.
Inflationary epoch A short period in the very early history of the universe immediately following the Big Bang, when the scale of the universe increased with unprecedented rapidity.
Intergalactic matter Material spread thinly between the galaxies. Evidently there is much more of it than used to be thought.
Interstellar matter Rarefied material spread between the stars.
Ion An atom which has lost one or more of its electrons, so that it has a positive charge.
Ion tail The gaseous tail of a comet. Unlike the dust tail,

it is usually almost straight.
IRAS The Infra-Red Astronomical Satellite, which operated throughout most of 1983.
Island universes An obsolete term for galaxies.
Julian day A count of the days, reckoning from 12 noon on 1 January 4713BC — a starting-point chosen by the mathematician Scaliger, who introduced it. (It has nothing to do with Julius Cæsar.)
Kepler's Laws Three important Laws of Planetary Motion, announced by Johannes Kepler between 1609 and 1618.
Kiloparsec One thousand parsecs (3,260 light-years).
Libration The apparent 'rocking' or 'tilting' of the Moon throughout the month, as seen from Earth. Because of librations, we can examine a total of 59 per cent of the total surface, though never more than 50 per cent at any one time.
Light-year The distance travelled by light in one year. Since light moves at 186,000 miles per second, one light-year is equal to about 5,880,000,000,000 or almost six million million miles.
Local group A group of galaxies of which our Galaxy is a member. Other members are M31 (the Andromeda Spiral), M33 (the Triangulum Spiral) and the two Clouds of Magellan.
Lunation The interval between one new moon and the next. It is equal to 29 days 12hr 44min. It is also known as the **Synodical month.**
Magellanic Clouds (Clouds of Magellan) The nearest important galaxies, regarded as companions of our Galaxy. Both are less than 200,000 light-years away. Unfortunately they are too far south in the sky to be seen from Britain.
Magnetosphere The area round a celestial body in which the magnetic field of that body is dominant.
Magnitude, Apparent The apparent brightness of a star (or other body) as seen from Earth. The brighter the object, the lower the magnitude.
Main sequence On the HR Diagram, most stars lie on a line from the upper left (hot, luminous, white or bluish stars) down to the lower right (dim red stars). This is the Main Sequence. The Sun is a typical Main Sequence star.
Mass The quantity of matter that a body contains. It is not the same as weight; thus an astronaut on the Moon has only one-sixth of his Earth weight, but his mass remains unaltered.
Megaparsec One million parsecs.
Meridian, Celestial The great circle on the celestial sphere which passes through the observer's zenith and both celestial poles. It therefore cuts the observer's horizon at the exact north and south points.
Messier numbers The numbers allotted by Charles Messier to the clusters and nebulæ in his famous catalogue, compiled in 1781.
Meteor A small particle, usually smaller than a grain of sand, which becomes luminous as it burns away in the Earth's upper air. Meteors are cometary débris, and are not associated with meteorites.
Meteorite A relatively large object, usually either made of iron or of stone, which enters the Earth's air and falls all the way to the surface. Meteorites probably come from the asteroid belt, and are not associated with meteors, though it is true that the term **meteoroid** is sometimes used to include both classes.
Micrometer A measuring device, used together with a telescope to measure very small angular distances — such as the separations of the components of double stars.
Micron One-thousandth of a millimetre (1/25,400 of an inch). There are 10,000 Ångströms to one micron.
Microwaves Radiations intermediate between infra-red and radio waves; wavelengths range from about 1 millimetre to 1 metre.
Midnight Sun The Sun as seen above the horizon at midnight.
Milky Way The luminous band stretching across the sky. In this direction we are looking along the main plane of the Galaxy, so that the Milky Way is merely a line of sight effect.
Minor planets A more correct name for asteroids.
Minute of arc One-sixtieth of a degree of arc.
Mira variable A long-period variable star. The name comes from the best-known member of the class, Mira Ceti in the constellation of the Whale.
Molecule A stable association of atoms. Thus a molecule of water, H_2O, consists of two atoms of hydrogen together with one atom of oxygen.
Multiple star A star made up of more than two physically associated components.
Nadir The point on the celestial sphere immediately below the observer.
Nano Abbreviation for 10^{-9} or one thousand millionth. Thus one nanosecond is one thousand millionth of a second.
Neap tide The tide produced when the Sun and Moon are at right angles to the Earth, so that they are, so to speak, pulling against each other.
Nebula A mass of dust and tenuous gas in space. If there are suitable stars in or near a nebula, the material shines; if not, it remains as a dark mass. Nebulæ are stellar birthplaces.
Neutrino A fundamental particle with no electrical charge and virtually no mass, so that it is extremely difficult to detect.
Neutron star The remnant of a very massive star which has exploded as a supernova. The protons and electrons run together to make neutrons, and the resulting density is very high.
Newtonian reflector The commonest type of reflecting telescope. The light is collected by the main mirror, and is diverted by a flat secondary mirror into the eyepiece.
Noctilucent clouds Strange clouds in the upper atmosphere, over 50 miles high. It is possible that they are due to dust left by meteors entering the air.
Nova The sudden flare-up of the white dwarf component of a binary system, which subsequently returns to its former obscurity.

Object-glass (Objective) The main lens of a refracting telescope.
Oort Cloud A 'cloud' of comets orbiting the Sun at a distance of about a light-year. It is named after J.H. Oort, who first suggested its existence. We cannot be sure of the reality of the Oort Cloud, but it does seem probable.
Opposition The position of a planet when it is exactly opposite to the Sun in the sky.
Optical double A double star in which the two components are not genuinely associated, but lie in almost the same line of sight as seen from the Earth.
Optical window The region of the electromagnetic spectrum through which radiations can pass without being blocked by the Earth's atmosphere. It extends from about 3000 to 9000 Ångströms (300 to 900 nanometres).
Orrery A model showing the Solar System, with the planets capable of being moved at their correct relative velocities round the Sun.
Panspermia theory A theory proposed by the Swedish scientist S. Arrhenius, according to which life was brought to the Earth by way of a meteorite.
Parallax The apparent angular shift of a distant body when observed from two different directions.
Parsec The distance at which a star would show a parallax of 1 second of arc. It is equal to 3.26 light-years, 206,265 astronomical units, or 19,150,000,000,000 miles. Actually, no star apart from the Sun is within 1 parsec of us.
Penumbra (a) The area of partial shadow to either side of the main cone of shadow cast by the Earth. (b) The outer part of a sunspot.
Perihelion The position in orbit of a planet or other body when at its nearest to the Sun.
Perigee The position in orbit of a body, moving round the Earth, when at its closest to the Earth.
Perturbations The disturbances in the orbit of a celestial body produced by the gravitational pulls of other bodies.
Phases The apparent changes of shape of a celestial body from new to full. Only the Moon, Mercury and Venus show complete phase cycles.
Photometer An instrument used to measure the intensity of light coming from one particular source.
Photon The smallest 'unit' or 'particle' of light.
Photosphere The bright surface of the Sun.
Planetarium An instrument used to show an artificial sky on the inside of a large dome.
Planetary nebula An old giant star which has thrown off its outer layers, and is therefore surrounded by a 'shell' of expanding gas. Planetary nebulæ are not true nebulæ, and are certainly not planets!
Poles, Celestial The north and south points of the celestial sphere.
Position angle The apparent direction of one celestial object with reference to another, from north through east, south and west back to north.
Precession The apparent slow movement of the celestial poles, which in turn affect the position of the celestial equator — and, hence, the right ascensions and declinations of stars.
Prominences, Solar Masses of glowing hydrogen above the Sun's surface. They were once, misleadingly, called Red Flames. With the naked eye they can be seen only during a total solar eclipse.
Proper motion The individual motion of a star on the celestial sphere. All proper motions are very slight, because the stars are so far away, and the constellation patterns do not alter noticeably over periods of many lifetimes.
Proton A fundamental particle with unit positive electrical charge.
Pulsar A rapidly varying radio source, now known to be a neutron star in quick rotation.
Quadrature The position of a Moon or planet in the sky when at right angles to the Sun as seen from Earth.
Quantum The amount of energy possessed by one photon of light.
Quasar Conventionally assumed to be a very remote, super-luminous object, probably the core of a very active galaxy.
Radial velocity The toward-or-away movement of a celestial body; positive if the body is receding, negative if it is approaching.
Radiant The point in the sky from which the meteors of any particular shower appear to come.
Radio galaxy A galaxy which is a very powerful emitter of radio waves.
Radio telescope An instrument used for collecting and analysing long-wavelength radiations from space. It is unlike an optical telescope, and does not produce a visible picture.
Radius vector An imaginary line joining the centre of a planet (or comet) to the centre of the Sun. According to Kepler's Laws, the radius vector sweeps out equal areas in equal times.
Reflection nebula A cloud of dust and gas which is illuminated by some nearby star.
Reflector An optical telescope in which the light from the target object is collected by a curved mirror or speculum.
Refraction The 'bending' or change of direction of a ray of light when passing through a transparent substance.
Refractor A telescope in which the light from the target object is collected by a lens known as an object-glass or objective.
Regolith The outermost 'loose' layer of the surface of the Moon or other planetary body.
Relative density The density of a substance compared with that of an equal volume of water. The Earth's relative density is 5·5. Also known as **specific gravity.**
Resolving power The ability of a telescope to separate objects which are close together. The larger the telescope, the better the resolving power.
Retrograde motion The revolution or rotation of a body in a sense opposite to that of the Earth. For example, Venus has retrograde rotation, since it spins on its axis from east to west.
Right ascension The angular distance of a body from

the First Point of Aries or vernal equinox, measured eastward. In practice it is usually given in hours, minutes and seconds of sidereal time.
Rill A linear depression on the Moon's surface.
Roche limit The distance from the centre of a planet within which a second body would be broken up by that planet's pull (assuming that the orbiting body is very 'loose').
Saros A period of 18 years 11·3 days, after which the Sun, Earth and Moon return to almost the same relative positions. This means that an eclipse is likely to be followed by another eclipse 18 years 11·3 days later. The Saros is not exact, but it was good enough to enable astronomers of Classical times to make reasonable eclipse predictions.
Satellite A secondary body moving round a planet.
Schmidt telescope A type of telescope which uses a spherical mirror together with a glass correcting plate at the top of the tube. It enables comparatively wide areas of the sky to be photographed on a single exposure.
Scintillation The official term for star-twinkling.
Second of arc One-sixtieth of a minute of arc.
Secular variable A star which is suspected of a permanent change of brightness in historic times.
Selenography The physical study of the Moon's surface.
Seyfert galaxies Galaxies with relatively small, bright nuclei and weak spiral arms. Many of them are powerful radio emitters.
Shadow bands Wavy lines seen across the Earth just before and just after totality during a solar eclipse. They are due to the Earth's atmosphere, and are not seen at every eclipse.
Shooting-star The popular name for a meteor.
Sidereal period The time taken for a planet or comet to complete one orbit of the Sun, or for a satellite to complete one orbit round its primary planet.
Sidereal time The local time reckoned according to the apparent rotation of the celestial sphere.
Solar parallax The trigonometrical parallax of the Sun: 8·79 seconds of arc.
Solar System The system made up of the Sun, planets, satellites, comets, meteoroids and interplanetary matter.
Solar wind A constant flow of electrified particles, streaming out from the Sun in all directions.
Solstices The times when the Sun is at its furthest from the celestial equator; declination 23½°N in northern midsummer (around 22 June) and 23½°S in southern midsummer (around 22 December).
Specific gravity The density of a substance compared with that of an equal volume of water. The Earth's specific gravity is 5·5. Also known as **Relative Density.**
Spectroscope An instrument used to split up the light from a luminous object. The production of a **spectrum** is carried out by means of a prism or some equivalent device.
Star A self-luminous gaseous body. The Sun is a typical star.
Sunspots Darker patches on the surface of the Sun. They are associated with strong magnetic fields, and appear dark only by contrast against the brilliant photosphere.
Supernova A tremendous stellar outburst. A Type I supernova is produced by the complete destruction of the white dwarf component of a binary system, while a Type II supernova is caused by the collapse of a very massive star. Neutron stars are the remnants of Type II supernovæ.
Synodic period The mean interval between successive oppositions of a planet.
Syzygy The position of the Moon in its orbit when new or full.
Tektites Small, glassy objects found in certain restricted areas of the Earth. They may or may not be meteoritic in nature.
Transit (a) The passage of a celestial body across the observer's meridian. (b) The passage of Mercury or Venus across the Sun's disk.
Troposphere The lowest part of the Earth's atmosphere, extending up to an average height of about 7 miles.
Ultra-violet radiation Electromagnetic radiation which is intermediate in wavelength between visible light and X-rays. It extends between about 4,000 Ångströms down to about 100 Ångströms.
Umbra (a) The dark central part of a sunspot. (b) The main cone of shadow cast by the Earth.
Van Allen Zones Zones round the Earth in which electrically charged particles are trapped by the Earth's magnetic field.
Variable stars Stars which change in brightness relatively quickly. They are of various types.
Vernal equinox (First Point of Aries) The point where the celestial equator is cut by the ecliptic, with the Sun moving from south to north. Because of precession, the vernal equinox has now moved out of Aries into the adjacent constellation of Pisces.
White dwarf A very small, very dense star which has exhausted its nuclear reserves. The most famous white dwarf is the Companion of Sirius.
X-rays Radiations in which the wavelength extends between 100 and 0.01 Ångströms, between the ultra-violet and the gamma-ray regions.
Year The time taken for the Earth to complete one orbit of the Sun: 365.26 days, or 365 days 6hr 9min 10sec.
Zenith The observer's overhead point: altitude 90°.
Zodiac A belt stretching right round the sky, to 8° either side of the ecliptic, in which the Sun, Moon and all planets apart from Pluto are always to be found. It passes through twelve constellations, and a thirteenth, Ophiuchus, actually crosses it between Scorpius and Sagittarius.
Zodiacal Light A cone of light rising from the horizon and stretching along the ecliptic. It is visible only when the Sun is a little way below the horizon. Its cause is interplanetary matter spread along the main plane of the Solar System. A still fainter extension along the ecliptic is known as the **Zodiacal Band.**

USEFUL ADDRESSES

The leading British observational society is the British Astronomical Association (Burlington House, Piccadilly, London W.1). Membership is open to all. Regular meetings are held, and there is a bi-monthly *Journal*

Most large towns have societies of their own. A full list of all local societies can be found in the annual *Yearbook of Astronomy*, published each autumn (Sidgwick and Jackson, London).

The British monthly periodical is *Astronomy Now*, published by Intra Press (193 Uxbridge Road, London W12 9RA). It is on sale at all major newsagents.

ACKNOWLEDGEMENTS

My grateful thanks are due to Paul Doherty, for his excellent illustrations, and to those who have been kind enough to allow me to use their photographs:

NASA,
The European Southern Observatory
Palomar Observatory,
Las Campanas Observatory,
South African Astronomical Observatory,
The Hatfield Polytechnic Observatory,
Bernard Abrams,
Dr Halton Arp,
John Fletcher,
Gerry Gerrard,
Peter Gill,
Commander Henry Hatfield,
Ludolf Meyer.

PATRICK MOORE
Selsey, Sussex

Index

Page numbers in *italic* denote illustrations

AAT (Anglo-Australian Telescope), 117
Achernar, 154, 156, 157
Achondrites, 38
Acrux, 156
Adams, J. C., 92
Adams, W. S., 126
Adrastea, 82
Aerolites, 38
Airy, G. B., 92
Akhenaten, 122
Albireo, 152
Alcor, 148,
Alcyone, 156
Aldebaran, 150 155
Aldrin, E., *51*, 52, 58
Alexander (Emperor of Russia), 27
Algenib, 134
Algol, 136
Alheka, 168
Alkalurops, 138
Al Nath, 135
Alpha Centauri, 11, 16, 156-7, 170
Alpheratz, 134
Alphonsus (lunar crater), 102
Altair, 152, 154, 155, 170
Amalthea, 82
Ananke, 82
Anaxagoras, 8
Andromeda, 135, *135*
Andromeda Galaxy, 13, 17, *17*, 142, 150-52, *151*, 154, 175, 181
Antarctica, 38, 117, 159
Antares, 13, 17
Antlia, 137
Aphrodite Terra, 68
Apollo missions, *51*, 52-3, *56, 58* 59,
Arcturus, 152
Arecibo, 121
Argo, 138
Ariel, 80, 89
Aristotle, 9, 36
Armagh Observatory, 174
Armagh Planetarium, 158
Armstrong, N., 52, 169
Arnold, H. J. P., 169
Arp, H., 183-4, 189
Arrhenius, S., 27, 41, 69, 70
Ashen Light, 68-9, *69*
Asteroids, 13, 62, 79, 80
Asterope, 156
Astræa, 80
Astrology, 16
Atacama Desert, 110
Atlantis, 10
Atlas (satellite of Sturn), 84
Atlas (star), 156
Atwood Globe, 158
Augustine, St., 8
Aumann, H., 161-163
Auriga, 135, 150
Auroræ, *19* 95-6, *95*, 98, 159
  noise from?, 96
Azelfafage, 138

BEMs (Bug-Eyed Monsters), 65
BN (Becklin-Neugebauer Object), 176
Baade, W., 181-2
Baily's Beads, 98
Barnard's Star, 164
Barringer, D, M., 31-2
Barrow, I., 46
Barwell Meteorite, 102
Bauersfeld, W., 158
Bayer, J., 136
Bean, A., 53
Beehive, *see* Præsepe
Beer, W., 57
Belinda, 88
Bellerophon, 134
Beringer, J., 193
Berossus, 138
Bessel, F. W., 9, 16, 126
Beta Lyræ, 177
Beta Pictoris, 163, *163*
Betelgeux, 13, 14, 17, 124, 125, 134, 136-7, 147
Bethe, H., 124
Bevis, J., 168
Bianca, 88
Biela's Comet, *105*, 106-7
Big Bang theory, 189
Billy, J. de., 46
Binoculars, uses of, 19-21, 25
Biot, J. B., 37
Birmingham, J., 46
Birr Castle, *120*, 172
Birt, W. R., 46
Black Drop, 98-9, 100
Black Dwarfs, 124
Black Holes, 129-30
Boksenberg, A., 190
Bondi, H., 189
Bonn, radio telescope at, 121
Boyden Observatory, 181
Brisbane, T., 140
Bruno, Giordano, 8
Busch, A., 158
Butterfly Nebula, 132

Calendars, eccentric, 73
Callisto, *79*, 81, 82
Caloris Basin, 65
Calypso, 84
Cancer, 152
Canis Major, 126, 137
Canis Minor, 137, 150
Canopsus Road, 118

Canopus, 9, 141, 155, 159
Capella, 150, 154, 170
Capricornus, 16
Carme, 82
Cassegrain reflectors, 20
Cassiopeia, 135-6, 148-9, 154
Cassini, G. D., 42
Cassini Division, 84
Cassini space-probe, 84
Castor, 150, 152, 170
Celæno, 156
Centaurus, 134, 155
Cepheid variables, 177-82
Cepheus, 135-6
Ceres, 80
Cernan, E., 53, 58
Cetus, 136
Challis, J., 92-3
Chambers, G. E., 143
Charon, 94
Chen, Y. N., 131
Chi Cygni, 136
Chiron, 80
Chimæra, the , 134
Chondrules, 38
Clark, A., 181
Cleomedes (lunar crater), 55
Clouds of Magellan, 157, 177, 181
Clube, V., 143
Coal Sack, 156, 176
Comets, 13, 28, 102, *103*, 169, 188
  Biela, 105-7
  Daylight, 104-5
  Encke, 28, 36, 38
  Halley, 102-6, 108-9
  lifetimes of, 36
  of 1882, 102
  Swift-Tuttle, 28
  Tempel-Tuttle, 28
  Thatcher, 28
Comte, A., 123
Conrad, C., 53
Constellations, 16
  Egyptian, 135
  legends of, 135-7
  list of, 144-5
  rejected, 137
Co-ordinates, celestial, 147
Copernicus (lunar crater), *55*, 60
Cordelia, 88
Corona Borealis, 152
Cosmic rays, 186
Cosmic Serpent, the, 143
Counterglow, *see* Gegenschein
Crab Nebula, *127*, 129, 133, 140, 167, 175
Creationism, 193
Cressida, 88
Crick, F., 169

61 Cygni, 11, 16
Cygnus, 152
Cyr, D.L., 72
Cyrillids, the, 33

D'Arrest, H., 92-3
Darquier, A., 132
Darwin's theory, 192-3
Davis, R., 118
Day, lengthening of, 190
Daylight Comet, 104-5
de Kövesligethy, Dr., 142
de Podmaniczky, Baroness, 142
de Vaucouleurs, G., 74
Deimos, 71
Delaunay, C., 93
Delta Pavonis, 165
Dendrochronology, 96
Deneb, 11, 13, 72, 152, 154, 155, 159
Denebola, 152
Desdemona, 88
Despina, 91
Dicke, R., 192
Dinosaurs, 143, 188
Dione, 84, 85
Dogon (tribe in Mali), 125
Doherty, P., 96
Doppler, C., 180
Doppler effect, 180-3
Douglass, A. E., 96
Drake, F., 168
Dubhe, 149
Duhalde, O., 137
Duke, C., 53
Dumbbell Nebula, *131*, 132

Eagle Nebula, 131
Earth, *6*, 62
  age of, 186
  life on, evolution of, 187-8
  shape of, 7, 8
  status of, 13
Eclipses, solar, 97-8, *107*, 109, *109*
Eisanga, E., 158
Elara, 82
Electra, 156
Electromagnetic spectrum, 160
Enceladus, 84, 85
Encke, J. F., 36, 92
Encke's Comet, 28, 36, 38
Epimetheus, 84
Epsilon Eridani, 163, 168-9
Equator, view from, 148
Eridanus, 154
Eros, 80
Escape velocity, 129

Eskimo Nebula, 132
Eta Carinæ, 140-1, *141*
Europa, 81, 82
Ezekiel, Book of, 142-3

FG Sagittæ, 132
Fabricius, D., 136
Fabricius, J., 136
Fallows, F., 140
Felis, 137
Fesen, R. A., 143
Field, G., 104
Fizeau, H., 180
Flamsteed, J., 104-6
Flood, Biblical, 143
Flying saucers, 40, 143
Fomalhaut, 155, 163
Fossil record, the, 192
Fraunhofer, J., 128
Frederik II, Duke, 158
Freud, S., 186
French Revolutionary Calendar, 73
Frost fairs (on the Thames), 96

Gagnon, Mr., 85
Galatea, 91
Galaxies, 13, 21, 181-2, 194
Galaxy, the, 8, 13, 86, *163*, 180
Galileo, 8
Galileo space-probe, 82, 85
Galle, J., 92,93
Gamma-rays, 160
Gamow, G., 124, 194
Ganymede, 81, 82
Garching (Munich), 116
Gaskin, Judge, 85
Gauss, K. F., 72
Gegenschein, the, 42
Gemini, *139*, 150
George III, King, 86
Gilbert, G. K., 31, 32
Gill, D., 102, 107
Gillett, F., 161
Giotto probe, 106
Gladstone, 10
Globus Ærostaticus, 137
Godwin, F., 50
Gold, T., 189
Goldstone, radio telescope, *171*
Gosse, E., 182
Gosse's Bluff, 33-4, *34-5*
Gottorp Globe, 158-9
Grand Tour, the, 76
Gravity-assist technique, 77
Great Wall of China, 169
Greeks, the, 8-9
Green Bank, radio telescope at, 168

Grimaldi (lunar crater), 47
Grootfontein (Hoba West) Meteorite, 36, 38, 39
Gruithuisen, F. von P., 57-8, 66, 68
Grus, 155
Guillemin, A., 136
Gully, L., 142
Gum, C., 138
Gum Nebula, 138

H-R (Hertzsprung-Russell) Diagrams, 122-6
Haggard, H. Rider, 9
Hale, G. E., 120
Halemaumau, 113
Hale Pohaku, 114
Hall, A., 71
Halley, E., 102, 104, 105, 108
Halley's Comet, 102-6, 108-9
Hamilton, A. J. S., 143
Hartwig, K., 142
Hawaii, 113
Heard, G., 72
Helene, 84
Helix Nebula, 132
Henbury Craters, 33, *33*
Henderson, T., 16
Heraclitus, 7
Hercules cluster, 11, 156, *173*
Herschel, C., 36, 38
Herschel, J., 53-7, 124, 141, 168
Herschel, W., 52, 86, 88, 175
Herstmonceux Castle, 116, 190-1, *191*
Hertzsprung, E., 124
Hidalgo, 80
Hilo, 113
Himalia, 82
Hind, J. R., 109
Hoba West meteorite (Grootfontein), 36, 38, *39*
Holwards, Phocylides, 136
Homestake Mine, 117-9
Hooke, R., 104
Hörbiger, H., 50
Horse's-Head Nebula, *173*
Housden, W. B., 74
Hoyle, F., 106, 169, 183, 189
Hubble, E., 116, 181, 183
Hubble Space Telescope, 121
Hululai volcano, 113
Humason, M., 181
Huth, 36
Hven (island), 57
Hyades, 150, 175
Hydra legend, 137, *138*
Hygeia, 80
Hyperion, 84, 85

IAU (International Astronomical Union), 134, 137
IRAS (Infra-red Astronomical Satellite), 161, *161, 162*
IUE (International Ultra-Violet Explorer), 160
Iapetus, 84, 85
Icarus, 80
Ice Ages, 123
Indian theories of the Earth, 8
Infra-red astronomy, 161
Ingalls, A. G., 18
Io, 81, 82
Irwin, J., 53
Isaac Newton Telescope, 116
Ishtar Terra, 68
Izar, 170

Jahangir, Emperor, 27
Jalandhar Meteorite, 27
James I, King, 54
Jansky, K., 166
Janus, 84
Jarrett, A., 181
Jeans, J., 137
Jefferson, President, 37
Jep, 54
Jewel Box cluster, 156
Jodrell Bank, 121, 166
Juliet, 88
Julius Cæsar, 46, *46*
Juno, 80
Jupiter, 13, 62, 64, 70, 76, 78, *100*, 100-1
  data, 82
  Great Red Spot on, 80, 82
  radio emissions from, 166

Kazantsev, A., 40
Keck Telescope, 121
Kepler, J., 47-9, 57
Kepler's Laws, 47, 48, 57
Kibaltchitch, 56
Kilauea volcano, 113
Kilburn, S., 73
Kirchhoff, G., 128
Kitt Peak, *119*, 168
Kohala volcano, 113
Kozyrev, N., 102
Kuiper, G., 115
Kuiper Airborne Observatory, *194*
Kulik, L., 37

LGM (Little Green Men) Theory, 167
L'Aigle meteorites, 37
La Palma, 116, 117, *121*
La Silla, 110-2, 121
Ladbroke's, odds given by, 167
Larissa, 91
Las Campanas, 137, 163
Lassell, W., 88
Le Verrier, U. J. J., 92
Leacock, S., 185
'League of Planets', 76
Leavitt, H., 177, 181
Leda, 82
Leeds Time, 73
Leo, 137, 150
Leonid meteors, 29, *29*
Lepus, 137
Librations, lunar, 48
Life elsewhere?, 160-71
Light, velocity of, 11, 129
Little Ice Age, 96
Littrow, C. von, 72
Local Group of galaxies, 181
Locke, R. A., 53-6
London Planetarium, 158
Los Muchachos, 116
Lovell, B., 166
Lovell Telescope, 121, *165*
Lowell, P., 73-4, 96, 181
Lucian, *True History* by, 44-5
Ludwig's Star, 152
Luna 3, 48, 60
Lunar Base?, 53
Lupus, supernova in, 140
Luther, Martin, 8
Lyra, 152
Lysithea, 82

Machina Electrica, 137
Mädler, J. H., 57
Magdeburg Experiment, 11
Magnetic storms, 90
Maia, 156
'Man in the Moon', 46
Manchurian aerolite, 78
Mariner 2, *67*, 70
Mariner 4, 71, 74
Mariner 10, 65
Marius, S., 152
Markab, 134
Mars, 13, 62-4, 70-5, *71, 75*
  atmosphere, 70
  calendar, 72
  canals, 73-5, *74*
  caps, polar, 71
  craters, 71-2
  dust-storms, 100-1
  life on?, 73
  pole star of, 72
  satellites, 71, 74, 75

Mauna Kea, 113-5, *114*, 161
Mauna Loa, 113-4
Maunder minimum, 96
McNaught, R., 137
Mecca, Sacred Stone at, 27
Méchain, P., 36
Medusa (mythological), 136
Menzel, D. H., 70
Merak, 149
Mercury, 13, 62, *63*, 64-5, 96
Merope, 156
Messier, 109, 168, 175
  Catalogue by, 175
Meteor Crater, 27-43, *30*
  age of, 29
  mining at, *30*, 31-2
  myths of, 32
  Trail down, 33
Meteorites, 38, 102
  list of large examples, 38
  numbers of, 38
  swords made from, 27
  types of, 38
Meteoroids, 13
Meteors, 28, 102
  showers, list of, 29
Metis, 82
Michanowsky, G., 138
Micrometeorites, 28
Microscopium, 137
Microwave background radiation, 192
Midnight Sun, 148
Mimas, 84
Mira, 136
Miranda, 87-9, *87*
Montanari, G., 136
Moon, the, *8, 10, 11*, 13, 44-61, *49, 50, 51*
  atmosphere, lack of, 60, 129
  composition, internal, 59
  craters, 22, 29, 59
  craters, list of, 60-1
  domes, 59
  eclipses, 9, *9*, 47
  far side of, 48, 60
  magnetic field, lack of, 60
  markings on, 44
  mountains, 59
  mountains, list of, 61
  origin, 59
  phases of, 10
  rays, 22, 60
  rills, 59
  rotation of, 48
  transient phenomena on (TLP), 101-2
  vegetation on?, 58
  view with binoculars, 22
  worship, 46
Moon Hoax, 53-7
Mount Wilson, 100in telescope at, 120, 175

NTT (New Technology Telescope), 111-2, *111*, 174, 177
Naiad, 91
Namibia, 36
Napier, W., 143
Nebulæ, 17, 21, 175-6
Neptune, 13, 62, 64, 76, 89-93
  data, 92
  Great Dark Spot on, 90, *90*
Nereid, 92
Neutrinos, 118-9
Neutron stars, 133
Newton, Isaac, 9, 102, 104, 128
Newtonian reflectors, 20, 25
Noctua, 138
Nordic Telescope, *121*
Northern Lights Planetarium (Tromsø), 158
Nuffield, Lord, 166
Nut (Egyptian goddess), 8

Oannes legend, 138
Oases, Martian, 73
Oberon, 88-9
Ocampo, S., 50, 52
Oleander, 158
Olympus Mons, *66*, 71
Omega Centauri, 156-7, *157*
Oort, J., 108
Oort Cloud of comets, 108, 188
Ophelia, 88
Orbiter probes, 48
Orgone energy, 186
Orientale, Mare, 101-2, *102*
Orion, 13, 14, *15*, 134, 137, 148-9, *148*, 152, *152*, 154-5
  Nebula in, 149, 175, 176
Orreries, 158
Owl Nebula, 132
Ozma, Project, 167

Painted Desert (Arizona), 67
Palitzsch, J., 109
Pallas, 80
Palomar reflector, 120, 175
Pandora, 84
Parallax, stellar, 16-7
Parkes Radio Astronomy Observatory, 117, 178, *178-9*
Pasiphaë, 82
Pegasus, 134, 155, 159
Pele (Hawaiian goddess), 113
Penzias, A., 192
Perrotin, H., 73
Perseid meteors, 28
Perseus cluster (Sword-handle), 135, *149*
Perseus legend, 135, *135*
Peter the Great, 105
Phæthon, 80
Phobos, 71
Phœbe, 84
Piazzi, G., 80
Pickering, W. H., 58
Pillar and claw mounting, *21*, 25
Pioneers 10 and 11, 79, 82, 169
Planetaria, 146, 158, *159*
Planetary nebulæ, *131*, 132
Planets, 13, 17
  of other stars, 164
  origin of, 64
  seen with binoculars, 22, 125
Plate tectonics, 113
Pleiades, the, 150, 156, 175
Pleione, 156
Pliny, 32, 143
Plutarch, 44
Pluto, 13, 62, 64, 93-4
Pointers, the, 149
Pole, view from, 146
Pole Star (Polaris), 11, 146, 148-9, 153, 170
Pollux, 150, 152
Pons, J. L., 36
Porta, A., 76
Portia, 88
Præsepe, *150*, 152, 156, 175
Procyon, 150
Prometheus, 84
Proteus, 91
Proxima Centauri, 124
Psalterium Georgianum, 137
Ptolemaic system, *7*, 57
Ptolemy, 124, 134, *134*
Puck, 88
Pulsars, 128, 133, *167*
Pup (Companion of Sirius), 126

Quadrans, 28, 137
Quadrantid meteors, 28, 137
Quasars, 182-4, *183*, *184*
Quatermass, Professor, 53

R Coronæ, 148
Radar, 160
Radcliffe Observatory, 117
Radio astronomy, 166
Radio telescopes, 121, 166

Reber, G., 166
Regulus, 152, 170
Reich, W., 186
Rhea, 84-5
Riccioli, G., 46-7
Rigel, 13, 14, 124, 134, 137, 148, 149, 159, 180
Ring Nebula, 132
Robinson, R., 174
Rogers, Mr., 85
Rømer, O., 158
Rosalind, 88
Rosse, Lord, 120, 168, 172, *172*
Rowley, J., 158
Rümker, C. L., 36
Russell, H. N., 124
Russian 236in reflector, 117, 120-1
Rutherford-Appleton Laboratory, 162
Ryle, M., 189

S Andromedæ, 142-3
Sadalbari, 138
Sagittarius, 153-4
Satellites, planetary, 13
Saturn, 13, 62, 64, 76, *81*, 83-5, 87
  data, 84
  rings, *83*, 85
  satellites, 84
  spots on, 84, 101, *101*
Sceptrum Brandenburgicum, 137
Scheat, 134
Schiaparelli, G. V., 73
Schmitt, H., 53
Schröter, J. H., 44, 52-3, 80
Schwabe, H., 96
Scorpius, *12*, 13, 137, 153-4
Scott, D., 53
Seneca, 124
Seven Sisters, *see* Pleiades
Shadow bands, 98
Shapley, H., 180-1
Shatter-cones, 35-7
Shepard, A., 53
Shooting-stars, *see* Meteors
Siberian impact, 36-8, 40, *40-1*
Sickle of Leo, 152
Siderites, 38
Siderolites, 38
Siding Spring, observatory at, 117, 178
Sigma Octantis, 153, 155, 159
Singapore, science centre at, 146, *147*
Sinope, 82
Sirius, 94, 124, *125*, 125-7, 141, 149, 155, 170
  Companion of, 126, 150
  redness of?, 124
Six pips, 190
Slipher, V. M., 181
Societies, astronomical, 26
Solar System, the, 13
Sombrero Hat Galaxy, 175
Southern Cross (Crux), 134, 154
Spica, 152
Spode's Law, 21
Spry, Reg, 24, 26
Star of Bethlehem, 142-3
Stars:
  brightest, list of, 153
  clusters of, 21, 156
  colours of, 17
  distances of, 9, 16
  double, 21, 52
  evolution of, 124-5
  luminosities of, 13
  magnitudes, 153
  movements of, 14
  names of, 138
  numbers of, 6
  spectra, 128
  twinkling of, 17, 112
  variable, 136
Steady-state theory, 189
Stephen, Judge, 16
Stornoway 'meteorite', 41
Struve, F. G. W., 16
Sun, 13, 129-30
  and radar, 160
  danger of observing, 25
  eclipses, 97-8, 107, 109
  future of, 130, 132-3
  magnetism, 153
  nature of, 8
  observing, 24-6
  origin of, 12, 21
  radio emissions from, 166
  worship, 122
Sunspots, *19*, 99
Supernovæ, 128, *129*, 133, 140, *140*, 168
Sutherland, observatory at, 117
Swift, Lewis, 28, 74
Symmes, J. C., 10
Syrtis Major, 70

Tarantula Nebula, *130*, 137
Tau Ceti, 165, 168, 169
Taurus, 135
Taygete, 156
Telescopes *23, 24*
  choosing, 25
  cost of, 18, 25
  largest, list of, 120
  magnification given by, 20, 25
  mountings, 24
  reflecting, 20, 25
  refracting, 20, 25
  tubes for, 24
  types of, 18
Telescopium, 137
Telesto, 84
Temple, R. K., 125
Tethys, 84
Thalassa, 91
Thatcher, A. E., 28
Thebe, 82
Thollon, M., 73
Tiberius, Emperor, 95
Time-dilation effect, 186
Time-travel, 185-6
Titan, 64, 76, 84
Titania, 88-9
Tombaugh, C., 94
Triton, 89, *91*, 92-3
Trojan asteroids, 80
Tsiolkovskii (lunar crater), *45*
47 Tucanæ, 156-7
Tunguskæ, *see* Siberian impact
Tuttle, H., 28
Tycho, 54, 57, 152
  (lunar crater) 60

Umbriel, 88, 89
Universe, evolution of, 189-92
  end of?, 194
Uranus, 13, 62, 64, 76, 87-9
  data, 88
  electroglow, 89
  poles, 69
  rings, *86*, 88
  rotation of, 89
  satellites of, 88
Ursa Major, *14*, 134, 148, 154, 155
Ursa Minor, 148
Ussher, J., 191

VLT (Very Large Telescope), 121
Van Allen zones, 96
van de Kamp, P., 164
van Maanen, A., 183
Vega, 16, 132, 152, 154, 155, 159, 161, 163, 180
Vela pulsar, 129, 138
Vela supernova, 138, 140
Venus, 13, 62, *62*, 64-71, 153
  transits of, 98-100
Vernal Equinox, 147
Verne, Jules, 185
Vesta, 80
Viking space-craft, 74

Virgil, 124
Virgo, 152
Vladivostok meteorite, 38
Volcanoes, nature of, 113
Voltaire, 75
von Braun, W., 11
von Däniken, E., 143
von Zach, F. X. 80
Voyager 2, 76-94, *77*, 169
Vredefort Ring, 35, *37*
Vulcan, 64, 96

Wegener, A., 113
Weigel Globe, 159
Weisberger, J., 50
Wells, H. G., 185
Whipple, F. L., 70
Whipple Observatory (Mount
 Hopkins), 121
Whirlpool Galaxy, 175
White Dwarfs, 125-6
Wickramasinghe, C., 106
Widmanstätten patterns, 38
Wilkins, H. P., 101
Wilkins, J., 50
William Herschel Telescope, 116
Windows, radio and optical, 164
Wilson, R., 192
Winslow (Arizona), 27
Wolf, M., 143
Wolf Creek Crater, *31*, 33
Wood, R. F., 131
Wren, C., 104

X, Planet, 94
X-rays, 160

Yerkes Observatory, 120
Young, J., 53

ZHR (Zenithal Hourly Rate), 28
Zodiac, the, 16
Zodiacal Light, 42-3, *42-3*
Zubenelchemale, 138